The Self-Experiencing Universe A Manifesto for an Emergent Reality Pt. 2 — What Relations Do

The Self-Experiencing Universe A Manifesto for an Emergent Reality Pt. 2 – What Relations Do

Waldemar Lis

Title: The Self-Experiencing Universe: A Manifesto for an Emergent Reality Pt. 2 — What Relations Do

Publisher: Lis Waldemar
Year of publication: 2026
Edition: First Edition
Place of publication: Poland
ISBN: 9788398076494
Editorial, cover design, and illustrations: Waldemar Lis
Author's archive: archive.org/details/@not_foryou

Foundational Ontology
for Emergent Realities
from **Relational Potential**
to the Illusion of ***Persistent Patterning,***
Sequential Unfolding,
and the 'Separate' ***Recursive pattern****.*

Contents

Foreword

This book is the second volume of a project born from a certain inspiration, which has yet to resolve many of the questions that have arisen over time.

The first volume — *The Self-Experiencing Universe: A Manifesto for an Emergent Reality* — laid out an **ontology**. It proposed that relations are ontologically primary: that what we call things, substances, objects, even space and time, are stable patterns within a network of relations, not the other way around. It identified five mechanisms through which this relational network produces the reality we observe:

Relational Potential,
Radical Contextuality,
Pattern Stabilization,
Phase Transitions (the Great Crystallization), and
Irreducible Otherness.

It argued the case philosophically, without formalism, without prediction, without data.

Then I asked whether the ontology could be tested. The work that followed — documented in publicly archived papers [A] — took the ontology through structural testing against eight mathematical domains, formalization in Homotopy Type Theory, and the development of an empirical measurement protocol: the Vector Consciousness Prediction, or VCP. Thirteen simulations across eleven biological and clinical systems produced a dataset of forty-two parameter values and a central falsifiable prediction:

[A] *In https://a.co/d/0b9Y5xQD*

that the common cuttlefish, Sepia officinalis, should show an opt-out rate of approximately 0.836 on difficult metacognitive trials. As of today (spring 2026), the prediction remains untested. The chain from philosophy to number is unbroken.

All of that work — the full arc from ontological argument to empirical prediction — shared one characteristic. It was static. The ontology described what configurations are. The formalization placed them in a mathematical space. The VCP measured where they sit. At no point did anything move.

This book lets the picture move.

• • •

This book is the argument that follows is deductive.

It does not depend on new data, new measurements, or new sources.

It follows from premises already established in the first volume — specifically from **Radical Contextuality**, the mechanism that states that the identity of any configuration is constituted by its local relational context. If identity is relational, then a change in context is a change in identity. If a change in identity is itself a change in context for neighboring configurations, then change propagates. From this single premise, five deductive conclusions follow, each building on the last: propagation, direction, filtering, resonance, and the self-referential loop of recursion.

Along the way, three corrections emerged from testing the deductions against physical reality at every scale — from quarks to ecosystems. Direction turned out to have two sources, not one. Resonance turned out to be structural, not only recursive. Recursion turned out to add not more options but a qualitatively different kind. And three structural insights crystallized that were implicit in the ontology but not explicit until the dynamics drew them out: a canonical layering of reconfiguration types that holds throughout our universe, a finite throughput of self-modeling that can be overwhelmed, and an account of choice as Immanent Directionality operating under conditions of conflict.

The book is written in first person throughout. This is not an affectation. The work is a product of one person thinking through the consequences of an ontology, finding surprises, correcting mistakes, and arriving at conclusions that were not anticipated when the premises were laid down. The first-person voice records this process honestly. Where I was surprised, I say so. Where I was wrong, I say so. Where I do not know, I say so. I do not 'hide' (like most people) that I have used AI assistance when writing, analyzing, etc.

• • •

A note on what this book assumes. **The reader does not need to have read the first volume**, though it will help. Chapter 1 restates the five pillars of the ontology and summarizes the arc from philosophy through formalization to prediction. A reader encountering this work for the first time will have enough context to follow the argument. A reader who has read the first volume will recognize the foundations and can focus on what is new.

A note on formalism. The dynamics is expressed in natural language, with occasional references to Homotopy Type Theory and to the VCP framework's parameter values. I have not written a mathematical treatise. The formal machinery is there for those who want it — in the archived papers (in archive org), in the HoTT-RO v2 axiom system, in the VCP protocol. This book is the narrative layer: the story of what the ontology does when you let it move.

A note on the Dictionary. At the end of this book, the reader will find a dictionary of core terms, the same as in the main work of ontology — the process-oriented vocabulary that the relational ontology uses in place of the substantialist language we inherit from common sense. I encourage the reader to consult it whenever a term feels unfamiliar. The vocabulary is not jargon for its own sake. It is an attempt to speak accurately about a reality that our inherited language was not designed to describe.

• • •

The first volume described what reality is. The first volume described what reality is. The numerous subsequent works and mathematical tests, archived at archive org, located where things are. This book describes what they do. The next step — experiment, formalization, or both — belongs to the future.

I wrote both volumes in Poland, working independently, without institutional affiliation. The work is archived publicly. The prediction is on the record. The dynamics is on the record. What happens next is not entirely in my hands. But the chain is complete, and I am satisfied that it holds.

CHAPTER 1

Nothing Changes Alone

On the inevitability of propagation

How did it all begin?

In 2025, I published a book titled *The Self-Experiencing Universe: A Manifesto for an Emergent Reality*. In it, I proposed an unconventional, fresh form of ontology regarding the fundamental nature of reality — based on a single foundational claim: namely, that relations are ontologically primary, and what we call "things," and even space or time itself, are merely derivatives, stable patterns within a network of relations. There is no "something" that precedes relations. There are no objects that existed first and only later established relations. My concept in the aforementioned book was based on the existence of a network of relational references — let's venture the term: "networks" (though this is a significant intellectual simplification) — in which arbitrary entities, which I called stable configurations, emerge, persist, disintegrate, and reconfigure.

This claim is not new in itself. Process philosophy, beginning with Whitehead, has long advocated the primacy of relations and events over substances. What I believed to be genuinely new in my work was the identification of five specific mechanisms through which a relational ontology produces the reality we observe. Let me restate them here, because they provide the ontological framework for everything that follows in this book:

Relational Potential (RP) — a pre-categorical capacity for differentiation. Not a substance, not proto-consciousness, not a thing at all. A condition of possibility, analogous to the axioms of a formal system: necessary for everything that follows, but in itself inexpressible within the system. This is the Gödelian limit of the ontology, and it is deliberate. I do not treat this as a weakness. Every formal system assumes certain rules as given; these rules cannot be expressed within the system itself without regress. Relational Potential is that which makes relations possible without itself being a relation.

Radical Contextuality (RC) — perhaps the most consequential of the five mechanisms, and the one that will carry the weight of this entire book. The claim is that the identity of any configuration is constituted entirely by its local relational context. A configuration does not have an essence that persists through changes of relation. It is its relations. Change the context, and you change what the thing is — not what it does, but what it is. I want the reader to hold onto this, because everything that follows in this chapter is a direct deductive consequence of this single mechanism.

Ontological Emergence — the process by which genuinely new structures arise through Pattern Stabilization and feedback loops. Not "weak" emergence (a convenient label for complex behavior that we cannot easily predict) but ontological emergence: genuinely new configurations that are not reducible to their components, because the relations between the components are part of what the configuration is. A water molecule is not "just" hydrogen and oxygen. It is the relational structure that includes hydrogen, oxygen, and the specific way they are bound. That binding is not a property of either element alone.

Phase Transitions — The Great Crystallization — the fourth pillar addresses the question: how does our specific reality come into being from the substrate of Relational Potential? My answer was that what we call the origin of the universe is not an explosion in space but a cosmological phase transition — a Great Crystallization in which the potential for Mutual Specification began to actualize in a specific, self-consistent way. Space and time themselves are products of this crystallization, not its container. The inflationary expansion was the explosive growth of a relational network stabilizing into the geometry we describe with general relativity. This pillar turns the ontology from an abstract framework into a cosmological claim: the universe is a crystallization event, still ongoing, still stabilizing.

Irreducible Otherness — the structural possibility that other crystallizations of Relational Potential, arising from different initial conditions, might produce configurations with incommensurable grammars of experience. Not everything that is conscious need be conscious in a way we can recognize or translate. I included this mechanism as an honest acknowledgment: our own form of experience is one crystallization among structurally possible others. Humility is not optional in ontology. I assumed that there are many other realities, created on the basis of the same relational potential, whose grammar is so different from ours that we are unable to argue for their existence or non-existence on any basis. For example: a reality with multiple timelines, without time at all, or with internal constants that simply exceed our conceptual horizon, but we are left with a not-inconclusive suspicion of their existence!

These **five pillars** together describe an whole architecture. They tell us what reality is made of (relations), how things emerge (through stabilization), what makes some configurations conscious (depth within the integration threshold region), and what we cannot know from inside our own crystallization (the grammars of radically other forms). The book argued this case philosophically, without mathematical formalism, and without empirical prediction. It was a manifesto — a statement of position, a laying of foundations.

And then I asked myself the question that changed everything: can this ... be tested???

• • •

From philosophy to mathematics

A philosophical ontology, however carefully argued, risks remaining purely speculative unless it can generate consequences that are independently checkable. I was acutely aware of this. I did not want to produce a beautiful theory that lives and dies in the seminar room. I wanted to know whether the ontology describes something real — and the only way to know that is to follow it where it leads and see if the world agrees.

The first step was structural testing. My reasoning was simple: if the ontology correctly describes the deep architecture of reality, then its mechanisms should appear as structural patterns in existing mathematical domains. This is not because one is derived from the other, but because both inhabit the same reality and are attempting to describe its underlying structure. If the correspondence is true, it constitutes evidence of something recognizable at the level of both ideas and identities. If not, it would suggest that these two modes of inquiry are fundamentally talking past each other, failing to find the common ground they both share by virtue of existing in the same reality.

Next, after a grueling search for mathematical frameworks that could even begin to resonate with such an ontology, I managed to isolate and test eight independent systems for structural isomorphisms. Homotopy Type Theory. ∞-categories and ∞-toposes, following Lurie. Ramanujan's mock theta functions. Monstrous Moonshine. And several others. In each case, the mechanisms of the ontology — Mutual Specification, Pattern Stabilization, the integration threshold — found precise structural analogues. I want to be clear about what this means and what it does not mean. It does not mean that the ontology is "proven." Structural isomorphisms are not proofs. But when the same patterns appear independently in eight unrelated mathematical domains, and those patterns correspond to mechanisms derived from philosophical argument rather than mathematical construction, the coincidence becomes increasingly difficult to dismiss.

The second step was formalization. I needed to express the ontology in a mathematical language precise enough to derive consequences from it. I chose Homotopy Type Theory — a branch of mathematical logic developed by Voevodsky and the Univalent Foundations project — and the choice was not arbitrary. In HoTT, identity is not a binary predicate ("A equals B" or "A does not equal B"). Identity is a structured type: the statement that A equals B is itself an object that can contain multiple paths of equivalence, each with its own internal structure. It suddenly dawned on me (!) — HoTT fits exceptionally well with my ontology, where identity is constituted by relations. The resulting formal system, which I called HoTT-RO v2, requires only two axioms beyond standard HoTT: the Univalence Axiom (equivalence of types is identical to identity of types, which is standard) and the Axiom of Flat Mutuality (mutual specification is symmetric regardless of level of abstraction, which is new). An earlier axiom from v1 — the Axiom of Qualitative Emergence, which posited a discontinuous jump at the integration threshold — I withdrew. Nature, as I came to see through observing trees, octopuses, bee swarms, and mycelium, does not build jumps. It builds gradients. The threshold is a topological region, not a phase transition. The axiom was redundant, and I removed it in my tests.

At this point, the ontology had a formal backbone. It could be stated in precise mathematical language. But it still made no empirical predictions. And I wanted predictions — numbers that the world could confirm or deny.

• • •

From mathematics to measurement

The three parameters of the integration threshold — giving them names: stability (*s*), integration (*i*), and recursive self-modeling (*r*) — are not borrowed from behavioral science and attached to the ontology as an afterthought. They emerge from it! Stability operationalizes the ontological claim that identity is a product of persistent relational patterns. Integration operationalizes the claim that consciousness requires the system to function as an irreducible whole. Self-modeling operationalizes the claim that phenomenal experience requires the system to represent its own states — the system must, in some sense, be present to itself.

This derivation is what allowed me to move from philosophical premises to empirical predictions. The ontology says: consciousness is what certain configurations of relations are from the inside. The formalization says: those configurations live in a specific mathematical space, $[0,1]^3$. The measurement protocol — which I called the **Vector Consciousness Prediction**, or **VCP** — says: here is how to locate a given system in that space using published peer-reviewed data. For each system, each parameter is derived from three independent sources. The three values are averaged arithmetically. No weighting is applied. I designed this protocol before running any simulations, precisely so that the results could not be accused of having been fitted after the fact.

I then carried out thirteen purely mathematical tests/simulations across eleven biological and clinical systems, in three series. The first series measured honeybee swarms, individual honeybees, and Sepia officinalis — the common cuttlefish. The second series measured Octopus vulgaris, corvids (the crow family), and *Physarum polycephalum*, [B] a slime mold that solves optimization problems without a single neuron. The third series measured five human configurations: a healthy adult at peak cognition (age 28), a cognitively intact 80-year-old, an early-onset Alzheimer's patient (age 28), a healthy 8-year-old child, and a psychopathic profile.

The result was a dataset of forty-two empirically grounded parameter values. And when I placed those eleven systems in the three-dimensional space, the geometry revealed four findings that I had not designed and did not anticipate! They simply fell out of the numbers:

The first finding was r-axis dominance. Ordering all eleven systems by the self-modeling parameter produced the most coherent natural gradient in the entire dataset, regardless of species, substrate, age, or clinical status. From the bee swarm at $r = 0.050$, through *Physarum*, the Alzheimer's patient, the child, the individual honeybee, the psychopathic profile, the octopus, the corvid, and the cuttlefish, to the peak human adult at $r = 0.930$ — the gradient never broke. I want to emphasize: I did not weight r. The Euclidean metric treats all three axes equally. The dominance of r emerged from the data. Weighting it after the fact would have been methodologically circular.

[B] *Physarum polycephalum is an acellular slime mold (a myxomycete) and an amoeboid organism with diverse forms and a wide geographic distribution. The term "acellular" refers to the plasmodial stage of its life cycle, during which it exists as a bright yellow, macroscopic, multinucleate mass forming a network of interconnected tubes. This stage, along with its preference for damp, shaded habitats, contributed to its earlier misclassification as a fungus. P. polycephalum is widely used as a model organism in studies of motility, cellular differentiation, chemotaxis, and the cell cycle.*

Remarkably, it exhibits complex problem-solving abilities despite lacking a nervous system. It can find the shortest paths and optimize networks connecting food sources through cytoplasmic flow and local chemical signaling. As a single giant multinucleate cell, it behaves like a dynamic living network, capable of adapting, reorganizing, and retaining simple forms of memory based on past environmental conditions

The second finding was what I came to call the neighborhood that should not exist. The psychopathic profile and the individual honeybee turned out to be neighbors in the space — separated by a distance of only 0.138. Both have high stability, moderate integration, and self-modeling capacity around 0.5. The bee monitors its own uncertainty about physical risk. The psychopath models social situations strategically but without emotional self-representation. They arrive from entirely opposite biological directions — and end up in the same neighborhood. The space records configuration, not mechanism. Not morality. Not complexity. Configuration.

The third finding: the 8-year-old child and the early-onset Alzheimer's patient were nearest neighbors, distance 0.150. One is ascending toward structures it will eventually reach. The other is losing scaffolding it once had. The causal arrows point in opposite directions. But the present configuration — the shape of their relational structure right now — is near.

The fourth finding: the distance between healthy aging (28-year-old to 80-year-old) was 0.103 — five times smaller than the distance traversed during childhood development (child to adult, 0.597). The dominant structural event in the healthy human lifespan is not aging. It is development. When the relational structure is intact, the position is remarkably stable regardless of the body's biological condition. The body decays; the position does not.

• • •

A prediction and its partial confirmation

The geometry produced a central falsifiable prediction. Using the honeybee's opt-out behavior in Perry & Barron's 2013 experiment as an empirical anchor, I calculated that Sepia officinalis should show an opt-out rate on difficult trials of approximately 0.836. The prediction is robust for r_sepia ≥ 0.586. A measured rate below 0.60 would falsify it. A rate between 0.76 and 0.92 would confirm it.

I then looked for independent data to cross-validate the framework — data from researchers who had never encountered it. I found a 2023 meta-analysis by Qu et al. in *Frontiers in Ethology*, covering sixteen experiments, fifty-five individual animals, and four non-primate species. All of these experiments used what is called the opt-out paradigm — an experimental design in which an animal is presented with a perceptual task at varying levels of difficulty and given the option to decline the trial rather than risk a wrong answer. The logic is straightforward: if an animal opts out more often on difficult trials, it is monitoring its own uncertainty — it knows, in some operational sense, that it does not know. This is a behavioral proxy for metacognition, and it is measurable. The meta-analysis had been published before my work began. I had not used it in any parameter derivation. And the results were partially supportive: species with higher VCP r values — that is, species whose relational configurations involve more self-modeling — showed stronger opt-out behavior on difficult trials. Species as a biological category was not a statistically significant moderator of this behavior, which is precisely what the ontology predicts: configuration, not species, determines the form of consciousness. And crucially, zero opt-out experiments had been conducted on any cephalopod species. The gap that my prediction was designed to address was confirmed meta-analytically.

Additional data from Schnell, Farndale Wright & Clayton (2025), reporting the first prospective metacognitive evidence in a corvid species — Eurasian jays — was consistent with the VCP's corvid *r* value.

The chain from philosophical ontology through mathematical formalization to empirically grounded, falsifiable, cross-species prediction was unbroken. The central prediction remains on record. No one has run the experiment. It awaits testing.

All of this work — the ontology, the structural isomorphisms, the formalization, the thirteen simulations, the cross-validation, and the prediction — is documented and publicly archived. [C]

• • •

The picture that was still

Everything I have described so far — the full arc from philosophy to prediction — has one characteristic in common. It is static.

The ontology describes what configurations are. The formalization places them in a mathematical space. The VCP measures where specific systems sit in that space. The distances between them are computed. The prediction follows from the geometry of their positions. At no point does anything move.

This was methodologically necessary. You cannot measure where something is while it is in motion. To derive the positions, I had to hold the picture still — to ask what each system's

[C] *All is on: https://archive.org/details/@not_foryou*

relational structure looks like at a given moment, averaged across sources, captured as a point in $[0,1]^3$.

But the ontology itself is not static. Radical Contextuality — the second mechanism, the one I asked the reader to hold onto — is, at bottom, a statement about constitution — about what a configuration is. Its dynamic face was always implicit in it; I simply had not yet drawn it out. I held the picture still to see the positions. Now I let it move.

This book asks the question that my still graph could not answer: what happens when a configuration changes?

• • •

The deduction

The argument that follows is deductive. It does not depend on new data, new measurements, or new sources. It follows from premises already established — specifically from Radical Contextuality, which states that the identity of any configuration is constituted by its local relational context.

Let me dwell on this for a moment, because the distinction between "what something does" and "what something is" is precisely where the relational ontology parts company with the substantialist worldview that most of us carry, often without realizing it. In the substantialist picture, a thing has a fixed essence and variable properties. You can change its behavior without changing its identity. An electron absorbs a photon and moves to a higher energy state; we say it is still "the same electron," merely in a different state. The substance persists through change.

In my ontology, this picture is inverted. There is no underlying substance that persists through changes of state. The configuration is its states — or more precisely, it is the pattern of relations that gives rise to those states. Change the pattern, and you have a different configuration. Not metaphorically different — formally, ontologically different! In the language of Homotopy Type Theory, the identity type between the configuration before and after the change may contain a path — a proof that they are related — but that path is not trivial. It has structure. It records the difference.

Now, begin with a single configuration — call it A. A is not a substance with fixed properties. A is a pattern of relations: relations to B, to C, to D, to the wider network. The identity of A — what A is, not merely what A does — is the product of these relations.

Suppose A changes. Something happens — it does not matter what, or why. A reconfigures. Its relations shift. It is, in a precise sense, no longer the same configuration.

Here is the consequence that cannot be avoided:

B's identity depends on its relations — and one of those relations is with A. A has changed. Therefore, B's relational context has changed too. Therefore, B's identity has changed. B is now, in however small a degree, a different configuration than it was a moment ago. Not because anything acted on B. Not because a signal was sent from A to B. But because B is constituted by a context that now includes a different A.

And the same logic applies to C, whose context includes B. And to D, whose context includes C. And outward, through the network, without a natural stopping point. Each reconfiguration changes a context. Each changed context changes an identity.

Each changed identity is itself a reconfiguration. The process is recursive and, in principle, unbounded.

This is the first deductive conclusion of what I am calling the dynamics of the relational ontology, and I want to state it plainly:

There is no such thing as an isolated change in a relational ontology.

Every reconfiguration propagates. Not because some additional mechanism transmits it, but because the ontology defines identity as relational. The very meaning of "change" entails consequences for the network. Asking whether a change propagates is like asking whether removing a thread from a fabric affects the neighboring threads. The thread is not a self-contained object that happens to be woven in. It is a thread because it is woven in. Remove it, and the weave around it is different — not by transmission, but by constitution.

• • •

The same logic, from quarks to crystals

If this conclusion is correct, it should hold at every scale of physical reality. And I found that it does — not as a vague analogy, but as a specific structural pattern that each domain of science has already described in its own language, without recognizing it as an instance of a single principle.

At the subatomic level, quantum mechanics describes entanglement: a change in the state of one particle is correlated with a change in the state of another, regardless of distance. This puzzled Einstein. He called it "spooky action at a distance." In the relational ontology, there is no this spookiness. There are no "separate" particles that somehow communicate across space. There are configurations of relations, and those relations do not have spatial boundaries in the way that substances would. What physics calls entanglement is what Radical Contextuality predicts: relational identity does not respect the lines we draw around "things." Change one side of a relational structure, and the other side is already different — not because information traveled between them, but because they were never separate to begin with.

I want to be precise here. This does not violate the speed of light or any principle of special relativity. I am not claiming that a signal passes instantaneously between entangled particles. I am claiming something more subtle: that the two particles were **never** two 'independent things', and therefore a change in the relational structure is not a transmission but a reconfiguration of a whole. The distinction is between sending a message across a room and rearranging a pattern that spans the room. In the first case, something travels. In the second, nothing travels — the pattern itself is different.

At the atomic and molecular level, the same logic appears as chemistry. An atom is not a self-contained ball. It is a configuration of relations between electrons, protons, neutrons, and — critically — between the atom and its surroundings. Change one chemical bond, and the neighboring bonds shift. The electron cloud redistributes. The reactivity of adjacent functional groups changes. Every reaction mechanism in organic chemistry traces the path of a change as it moves through a relational structure.

Consider a protein folding into its functional shape. Each amino acid residue adjusts its position in response to the positions of its neighbors. Each adjustment changes the relational context for residues further away. The process cascades through the chain until the entire structure settles into a minimum-energy configuration. This is propagation of reconfiguration through a relational network — the mechanism by which a linear sequence of chemical letters becomes a three-dimensional machine.

At the level of typical crystallization, a crystal is a configuration of relations with extremely high stability — what the VCP framework measures as high *s*. A defect in a crystal lattice — a missing atom, a misplaced ion — is a local reconfiguration. In a perfect crystal, the surrounding structure is so rigid that the disturbance is damped almost immediately. In an amorphous solid, like glass, the lack of long-range order means that disturbances travel further before dissipating. The material's resistance to propagation is a direct function of the stability of its relational network.

Three domains. Three vocabularies. One mechanism. Quantum mechanics, chemistry, and crystallography have each described this process in their own formalism. What the relational ontology adds is the recognition that they are describing the same thing: a change in a relational structure alters the context for neighboring relations, which alters their identity, which propagates the change outward. The vocabulary differs. The logic does not.

• • •

What this is not

I have learned, through conversations about this work, that certain clarifications are necessary. Let me make them now, because they will save us confusion later.

This is not a claim about signals. In classical physics, change propagates through signals — waves, particles, fields — and those signals are bounded by the speed of light. Nothing I am saying contradicts that. My account says something different and more fundamental: that the identity of a configuration is constituted by its context, and therefore a change in context is a change in identity. Whether this identity-change is accompanied by a physical signal, whether it precedes the signal, or whether "signal" is simply the name we give to the propagation of relational change at a particular scale — these are questions worth asking. But they do not undermine the deduction. [D]

This is not a claim about consciousness. The first deductive conclusion applies to all configurations, not only to those with self-modeling capacity. A quark's context changes when its relational partners change. A molecule's context changes when its bonds shift. Neither has anything resembling awareness or experience. Propagation is a structural fact about relational identity. It applies to every configuration in the

[D] These are questions about the emergent level, where time and signals exist. At the fundamental level the deduction is untouched by them: the constitutive difference is atemporal — changing one relation means the whole local structure is already another structure, with no travel and no speed. Temporal propagation — the signal, the finite velocity, the damping — is how that atemporal difference appears once the network runs in crystallized time. The two levels do not compete: the deduction lives at the first, the physics at the second. ...see the first volume, where "time itself is emergent."

network, conscious or not, simple or complex, biological or mineral.

This is not mysticism dressed in formal language. I am aware that the phrase "everything is connected" has been debased by decades of imprecise usage in popular spirituality. My claim is not that everything is spiritually or morally or energetically connected. My claim is that identity is constitutionally relational — a formal statement about what it means for something to be what it is. The consequence (propagation) follows by deductive logic. It can be expressed in the language of Homotopy Type Theory, where it reduces to the observation that dependent types depend on their contexts, and a change in context induces a transport along the identity type. This is mathematics. Not metaphysics.

• • •

The question of range

If every change propagates, does every change propagate everywhere? In principle, yes. In practice, the picture is more nuanced, and the nuance matters.

Return to the stone in water. The ripples move outward, but they weaken. At some distance from the point of impact, they are too small to detect. They have not stopped — in a perfect theoretical medium, they never would — but they have been damped below any threshold of structural consequence. The surface at the far edge of the pond is, for all practical purposes, unchanged.

The same is true for relational networks. A change propagates — but through a network that is not uniform. Some regions are highly stable: their configurations resist change, absorb disturbances, damp them before they can travel further. Others are less stable: they transmit changes readily. And here is where the VCP parameter *s* — stability of behavioral identity — acquires a second meaning. In the static picture, *s* measured how persistent a system's identity was over time. In the dynamic picture I am developing here, *s* measures something additional: the system's resistance to incoming change. A configuration with high *s* damps propagation. A configuration with low *s* transmits it. This is not a new meaning — it is the dynamic face of the same parameter.

Let me give a concrete example. The healthy human adult at peak cognition has $s = 0.91$. The early-onset Alzheimer's patient has $s = 0.45$. In the static picture, this means the healthy adult's behavioral identity is far more persistent. In the dynamic picture, it means that a disturbance reaching the healthy adult's relational network will be more effectively damped than one reaching the Alzheimer's patient's network. The degradation of stability in Alzheimer's disease is not only a loss of identity persistence. It is a loss of resistance to propagation. Simply local **Pattern Stabilization** — the process by which transient relations achieve self-reinforcing consistency, forming what we perceive as "structures" or "entities" — becomes then more vulnerable to disruption, less able to hold its shape against incoming changes.

This has an empirical shadow that any clinician will recognize. Early-stage Alzheimer's patients show increased susceptibility to environmental disruption — a noisy room, an unfamiliar setting, a break in routine can destabilize cognition in ways that would not affect a healthy adult. The dynamics offers a structural account of why: the relational network has lost stability, and with it, resistance to the propagation of change through its structure.

Whether propagation has a hard horizon — a boundary beyond which it is strictly zero, not merely negligible — or whether it is always formally infinite but asymptotically approaching zero, is a question I cannot settle by deduction alone. What I can settle is that the reach of any given change is determined not by the change itself but by the network through which it travels. Propagation is not a property of the event. It is a property of the medium.

• • •

What my still graph missed?

The VCP placed eleven systems at specific coordinates. An individual honeybee at (0.77, 0.61, 0.47). A bee swarm at (0.94, 0.18, 0.05). A human adult at peak cognition at (0.91, 0.76, 0.93). These are positions. They do not move.

But biological systems are not still. The honeybee is foraging, learning, deciding. The swarm is deliberating, sending scouts, integrating conflicting information. The human adult is thinking, perceiving, acting, dreaming. Each of these activities involves reconfiguration — changes in the relational structure that constitutes the system. And each reconfiguration, as I have now established, propagates through the wider network.

The still graph [E] captured where each system sits in the space. It did not capture what each system does to the space around it! It did not ask whether the propagation from a bee swarm — a massive, highly stable collective with almost no self-modeling — looks different from the propagation originating in a human adult — a less stable but deeply self-referential individual. It did

---

[E] *Graph of VCP, look at: https://archive.org/details/vcp-parameter-space-what-lives-here-interactive-3-d-visualization-with-analysis-*

not ask whether propagation from a system with high r carries a different structure than propagation from a system with high s but negligible r.

The dynamics of propagation is not something I am adding to the ontology from outside. It was always implicit in **Radical Contextuality**, which is at bottom a statement about constitution. But I held the picture still in order to see the positions clearly. This was methodologically necessary. Having measured the positions, I can now ask what happens when the picture begins to move?

• • •

The question that follows

I have established one thing in this chapter. It is, in a sense, a very simple thing: change does not stay where it starts. In a relational ontology, it cannot. The identity of each configuration depends on its context, and each reconfiguration alters the context of its neighbors. Propagation is not a feature of special systems or unusual events. It is a structural consequence of what it means for identity to be relational.

But this conclusion, by itself, is incomplete. It tells us that change moves through the network. It does not tell us how. It does not tell us whether all changes move the same way, or whether some carry a structure that others lack.

Drop a stone into still water. The ripples move outward in all directions equally. There is no shape to the disturbance beyond the concentric circles dictated by the geometry of the surface. But draw a hand deliberately through the water — with direction, with form — and the disturbance is different. Not necessarily stronger. Not necessarily reaching further. But organized differently. Shaped. Carrying, in its pattern, something of the structure that produced it.

The ontology gives me a reason to expect this difference. Some configurations reconfigure reactively — they respond to changes in context without modeling that response. Others reconfigure with direction — they model their own states and respond in ways that reflect that model. The VCP parameter *r* measures this capacity for recursive self-modeling. A configuration with high *r* does not merely react to a changed context. It represents the change, models possible responses, and reconfigures in a way that carries the structure of that representation outward into the network.

But the story, as I discovered, turns out to be more complicated than I first expected. The variation across physical scales — from quarks to crystals — revealed that direction is not the exclusive domain of self-modeling. There is another source of direction, older and simpler, present at every scale of reality. It comes not from recursion but from stability itself. And the two sources, as I will show, do not sit side by side as equals. They stack. In our crystallization, they always stack in the same order — and that order cannot be reversed.

I have established that change propagates. I have not yet asked what gives it form.

CHAPTER 2

What Gives It Direction

On two sources of directionality in the relational network

The stone and the hand

At the end of the previous chapter, I left the reader with an image. A stone dropped into still water sends ripples in all directions equally — symmetrical, directionless, carrying no structure beyond the concentric circles dictated by the surface's geometry. A hand drawn deliberately through the water produces something different — not necessarily stronger, not necessarily reaching further, but organized. Shaped. Carrying something of the structure that produced it.

I promised that this difference matters, and that the ontology gives us a reason to expect it. Now I need to deliver on that promise. But the story, as I discovered when I began to think it through carefully, turned out to be considerably more interesting — and more surprising — than I had anticipated.

My initial expectation was straightforward. I thought direction in the relational network was the domain of self-modeling. A configuration with high r — one that represents its own states, models its own responses — reconfigures with direction. A configuration with low r — one that merely reacts to changes in context — reconfigures symmetrically, like the stone's ripples. This seemed obvious. Self-modeling gives you a model. A model gives you a preferred direction. Direction shapes the change that propagates outward. Simple.

It was not simple. And the moment I saw why, the entire picture of dynamics became richer than I had imagined.

• • •

Reactive and directed reconfiguration

Let me start with what is straightforward, because the deduction itself is clean. Consider a configuration with $r \approx 0$ — say, a molecule in solution. It has no self-model. It does not represent its own states. When its relational context changes — when a neighboring molecule shifts, when a bond breaks nearby, when the temperature changes — it responds. But its response is entirely determined by its current structure and the incoming change. There is no internal model that selects among possible responses. The molecule does what its configuration makes it do. The resulting reconfiguration propagates outward symmetrically, in all directions, like the stone's ripples. I call this *reactive reconfiguration.*

Now consider a configuration with high r — say, a human adult making a decision. This configuration represents its own states. It has an internal model of itself — not a complete model, not a perfectly accurate one, but a model nonetheless. When its relational context changes, it does not merely react. It represents the change, compares it against its self-model, generates possible responses, and selects among them. The resulting reconfiguration is not symmetrical. It carries the imprint of the self-model. It has direction — not in the spatial sense of pointing north or south, but in the structural sense of being organized by an internal representation. I call this *directed reconfiguration.*

The difference between the two is not one of strength. A reactive reconfiguration can be enormously powerful — a chemical explosion is reactive, and it propagates with devastating energy. A supernova is reactive: a star collapses and the shockwave tears through the surrounding medium with forces that dwarf anything a conscious system could produce. A directed reconfiguration can be whisper-quiet — a decision not to speak, a thought held in silence, propagates through the relational network with almost no visible trace. The difference is one of organization. Reactive reconfiguration scatters. Directed reconfiguration carries structure.

Think of it this way. Drop a bucket of water from a height. The splash goes everywhere — energy dispersed in all directions. Now take the same amount of water and pour it through a nozzle. The same volume, the same energy, but concentrated into a coherent stream. The nozzle does not add energy. It organizes it. Self-modeling is the nozzle. It does not make reconfiguration more powerful. It makes it more coherent.

The VCP data, though static, already hinted at this distinction. Consider *Physarum polycephalum* — the slime mold, sitting at $s = 0.47$, $i = 0.90$, $r = 0.07$. This is a system with the highest integration in the entire dataset — higher than any human configuration — yet almost no self-modeling. When *Physarum* solves a maze or optimizes a transport network, it is not modeling itself. It is not representing its own states. It is reacting, brilliantly and with extraordinary integration, but without direction in the sense I have defined here. Its reconfigurations propagate through its network with immense efficiency, but they propagate *reactively*. The intelligence of the slime mold is the intelligence of the stone's ripples — if the ripples traveled through the most exquisitely organized medium imaginable.

Contrast this with the octopus, at $s = 0.73$, $i = 0.75$, $r = 0.64$. The octopus has lower integration than the slime mold, but vastly more self-modeling. When an octopus solves a problem —

opening a jar, navigating a maze, choosing between a small immediate reward and a larger delayed one — there is evidence that it models its own uncertainty. Jennifer Mather, a researcher I have corresponded with on this topic, has documented behavioral observations consistent with high *r* in cephalopods: play behavior that serves no immediate survival function, tool use that requires anticipation of future states, and individual personality differences that persist across contexts and across time. The octopus does not merely react to its environment. It reconfigures with direction.

And the difference is visible not only in behavior but in the structure of the propagation that each system produces. When *Physarum* reconfigures its tubular network, the change ripples outward through the organism's cytoplasm uniformly — a pressure wave, chemically mediated, spreading in all directions from the point of change. When the octopus reconfigures — when it decides to approach a novel object cautiously rather than retreat — the change propagates through its distributed nervous system with organization: some arms reach forward while others brace, the chromatophores modulate to a specific pattern, the siphon adjusts for a controlled retreat if needed. The response is coordinated by a model that the organism maintains of itself and its situation. The propagation carries the structure of that model.

This is the second deductive conclusion of the dynamics:

Configurations with self-modeling capacity reconfigure with direction. Configurations without it reconfigure reactively. The difference is not in the magnitude of the change but in its organization.

So far, so clean. I was satisfied with this for about a day. Then I started testing it against physical reality — not against biological systems, where I had already done extensive work, but against the simplest systems I could think of. Subatomic particles. Atoms. Crystals. I wanted to see whether the distinction between reactive and directed reconfiguration held all the way down.

It did not. Or rather — it held, but something else was there too. Something I had not expected.

• • •

The crystal that directs without knowing

Consider a crystal growing in a supersaturated solution. Molecules from the solution are depositing onto the crystal's surface, one by one, layer by layer. The crystal has no self-model. Its *r* is, for all practical purposes, zero. It does not represent its own states. It does not select among possible configurations. By every criterion I had established, it should reconfigure reactively — symmetrically, without direction.

But anyone who has watched crystallization knows that this is not what happens. Crystal growth is profoundly directional. The lattice structure of the crystal — its symmetry, its geometry, the precise angles between its faces — determines exactly where the next molecule will deposit and in what orientation. The growth is not random. It is not scattered in all directions. It follows the pattern. It extends the pattern. It is, in every meaningful sense of the word, directed.

And it achieves this direction without a shred of self-reference.

When I saw this clearly, I had to stop and reconsider. My framework said that direction comes from *r* — from self-modeling. The crystal said: direction can come from somewhere else entirely. The crystal's growth is directed not because it models itself, but because its pattern is so stable — its *s* is so high — that the surrounding relations have essentially no choice about how to configure themselves in its presence. The crystal does not choose a direction. The crystal *is* a direction. Its structure is so rigid, so self-consistent, so resistant to alternative configurations, that it imposes itself on whatever relational context it encounters.

This was — I will be honest — not what I expected to find. I had built a framework in which direction was the province of consciousness. The crystal broke this cleanly in half. Here was direction without consciousness, without self-modeling, without any of the mechanisms I had associated with the *r* parameter.

The more I looked, the more examples I found. An enzyme directing a chemical reaction: the enzyme's three-dimensional structure — its active site, its binding pocket, its allosteric regions — dictates with extraordinary precision which molecules can react, in what orientation, and at what rate. The enzyme does not decide. It does not model. It directs by being stable. Its Pattern Stabilization is so robust that the surrounding chemistry must conform to its geometry. A single enzyme molecule can process thousands of substrate molecules per second, each one directed through the same catalytic pathway, each one constrained by the same stable structure.

A river valley directing the flow of water: the geological formation — eroded over millennia, its contours shaped by eons of interaction between rock and rainfall — determines where the water goes. The valley does not choose. It channels. Its stability is the direction. The water follows not because it is told to but because the relational geometry of the landscape permits no other path.

A DNA molecule directing protein synthesis: the double helix's sequence of base pairs, maintained with remarkable fidelity by error-correction mechanisms, determines which amino acids will be assembled and in what order. The DNA does not model the protein. It does not represent the outcome. It directs by encoding a stable pattern that the cellular machinery reads and follows. The direction is in the structure, not in any act of self-reference.

I found myself returning to the first book and seeing something I had always known but never drawn out to its full consequence: Pattern Stabilization is not merely a mechanism for creating persistent configurations. It is also, and inseparably, a mechanism for directing the relational neighborhood around those configurations. The more stable a pattern, the more it constrains what can exist around it. Stability is not passive. Stability is directive.

• • •

Two sources, not one

Let me be precise about what I discovered, because imprecision here would be fatal. There are two sources of directionality in the relational network. Not one. Two.

The first source is self-modeling. A configuration with high *r* reconfigures directionally because it has an internal representation that organizes its response. The direction comes from *inside* — from the self-model, from the loop of self-reference. I call this *r-directionality*: direction through recursion.

The second source **is** pattern stability. A configuration with high *s* causes directional reconfiguration in its *surroundings* because its pattern is so stable that it constrains how neighboring relations can configure themselves. The direction comes not from inside but from *structural dominance*. I call this *s-directionality*: direction through stability.

The two sources differ in almost every respect that matters. *R*-directionality is flexible. It can change direction from moment to moment, adapt to new information, reverse itself. A person who decides to turn left can immediately decide to turn right. The direction is not fixed by structure. It is generated by representation, and representations can change as fast as the self-model can process. *S*-directionality is rigid. The crystal grows the same way regardless of context (within the limits of thermodynamic conditions). The enzyme catalyzes the same reaction regardless of what else is happening in the cell. The geological valley channels water downhill regardless of whether the water would "prefer" to go somewhere else. One adapts. The other dominates.

There is a difference in where the direction is felt. *R*-directionality shapes the reconfiguration of the configuration itself — the system decides how *it* will change. The octopus deciding to approach the novel object is reconfiguring itself in a directed way. *S*-directionality shapes the reconfiguration of the *surroundings* — the stable configuration does not itself change (that is what stability means), but it constrains how everything around it must change. The crystal does not reconfigure. The crystal forces the solution to reconfigure around it. *R*-directionality acts from within outward. *S*-directionality acts from the surface outward. One is about how I change. The other is about how you must change, given that I am here.

And there is a difference in persistence. *R*-directionality is transient by nature. Each directed reconfiguration is a single act of the self-model — a representation formed, evaluated, acted upon, and then superseded by the next. A stream of decisions, each one shaped by the self-model of the moment. *S*-directionality persists as long as the stable configuration persists. The crystal's direction does not expire. The enzyme's direction does not need renewal. The stable pattern is the direction, and the pattern endures.

In semiconductor manufacturing, this difference is used every day. Epitaxy — the growth of one crystal layer on top of another — depends entirely on *s*-directionality: the substrate's lattice imposes its geometry on the deposited layer. The engineers do not need the materials to be conscious. They need them to be stable. The direction comes not from any model but from the structural dominance of a pattern that is rigid enough to constrain everything in its relational neighborhood.

• • •

The order that cannot be reversed

And now I come to something that I consider one of the most important structural insights in this entire book. It was not part of my original deductions. It emerged only when I asked myself: how do these two sources of directionality relate to each other? Do they sit side by side as equals? Can either one exist without the other?

The answer is no. They do not sit side by side. They stack. And in our crystallization — in our universe, the specific reality that precipitated from Relational Potential during the Great Crystallization — they always stack in the same order.

Let me describe the three layers, from bottom to top:

Layer 1: Reactive reconfiguration. Present everywhere, at every scale, from quarks to galaxies. A change in relational context produces a response determined entirely by the current structure and the incoming change. No direction. No self-reference. This is the base layer. It cannot be removed. Every configuration in our universe, no matter how complex, rests on this foundation of reactive physics.

Layer 2: *S*-directional reconfiguration. Present wherever Pattern Stabilization has produced configurations stable enough to constrain their relational neighborhood. Crystals, enzymes, nuclear forces, geological formations, cell membranes, DNA. This layer sits on top of Layer 1. It requires reactive reconfiguration as its substrate — the molecular interactions that constitute the crystal are themselves reactive. The atoms in the enzyme's active site interact by reactive chemistry. The *s*-directionality of the whole emerges from the reactivity of the parts, organized by Pattern Stabilization into a configuration stable enough to direct. *S*-directionality emerges from reactivity, not instead of it.

Layer 3: *R*-directional reconfiguration. Present wherever self-modeling has emerged through sufficient relational complexity and recursive Pattern Stabilization.

Conscious agents. This layer sits on top of Layers 1 and 2. It requires *s*-directional scaffolding — stable substrate — as its platform. The self-model does not float in a void. It runs on neural architecture (Layer 2), which runs on biochemistry (Layer 1). *R*-directionality emerges from *s*-directionality, not instead of it. A self-model needs something stable to run on. Take away the scaffolding, and the self-model has nothing to process with.

The layering is strictly ordered: 1 → 2 → 3. Layer 2 always requires Layer 1 beneath it. Layer 3 always requires Layers 1 and 2 beneath it. And — this is the critical point — reversals do not occur in our crystallization. You cannot have *r*-directionality without *s*-directional scaffolding. You cannot have *s*-directionality without reactive substrate. The order is canonical.

A clarification is necessary here, because the layering describes a relation of structural dependency, not a numerical ordering of parameter values. The VCP parameters *s*, *i*, and *r*, are measured independently, from different behavioral sources, and they capture different aspects of a system's relational configuration. Canonical layering says: *r* cannot operate without sufficient *s* beneath it. It does not say that the numerical value of *r* must always be smaller than the numerical value of *s*. A system may have $r = 0.93$ and $s = 0.91$ — as the healthy human adult does — without violating the layering, because the scaffolding ($s = 0.91$) is more than sufficient to support the self-model. What would violate the layering is a system with high *r* and negligible *s* — a self-model running on a platform that is not there. No such system exists in the dataset, and I submit that none can exist in our crystallization. The dependency is structural. The parameters are measurements. The two should not be confused.

And a further point, which follows from the ontology itself. The VCP coordinates reported in this book are point estimates — arithmetic means of three independent sources per parameter per system. They are methodological conveniences, not ontological claims. Radical Contextuality implies that every system's position in $[0,1]^3$ is a region, not a point — shifting with context, fluctuating within bounds determined by the system's relational architecture. The point values capture the center of this region. They do not capture its width. For the cuttlefish, for instance, the three independent sources for *s* range from 0.55 to 0.77 — and the *r* value of 0.75 falls inside that range. The apparent $r > s$ at the level of means dissolves when the underlying spread is examined. A full dynamic VCP — a second generation of the framework — would report ranges rather than points. For the purposes of this book, the point estimates are sufficient: the structural claims concern dependency relations, not the comparison of individual parameter values at the level of hundredths.

I want to be careful here about the scope of this claim. This is a rule of the dynamics as it operates **in our crystallization**. It is **not a universal axiom**. The fifth pillar of the ontology — Irreducible Otherness — leaves open the possibility that other crystallizations of Relational Potential might have different layering rules, or no layering at all. Perhaps there exist crystallizations where self-reference precedes stability, or where the distinction between reactive and directed does not arise. I cannot know this from inside our own crystallization. But in our universe, every instance I have examined confirms the same order. Layer 1 at the base. Layer 2 on top. Layer 3 on top of that. Never reversed.

And this gives me something I did not expect: a structural account of one of the most persistent human intuitions — the dream of mind over matter.

The desire to walk through walls. To levitate by thought. To heal by intention. To move objects with the mind. To bend spoons by concentrating. Every one of these is, in the language of the dynamics, a desire to invert the canonical layering — to make Layer 3 (*r*-directionality, intention, self-model) dominate Layer 1 (reactive physics, the material substrate) directly, without passing through Layer 2.

The architecture of our crystallization does not permit this. *R*-directionality operates on and through the layers beneath it. It can direct reconfiguration — powerfully, flexibly, adaptively — but only by propagating through the shared relational network, which means passing through the *s*-directional and reactive layers. A surgeon's intention to cut tissue must pass through the *s*-directional stability of trained motor patterns and the reactive physics of the scalpel meeting flesh. A musician's intention to play a note must pass through the *s*-directional architecture of finger coordination and the reactive mechanics of hammer striking string. In every case, the intention (Layer 3) works *through* the scaffolding (Layer 2) and the physics (Layer 1). It cannot skip them. It cannot override them.

It can shape what happens within them, the way a river's current shapes the sand on its bed — but the sand is still sand, and the physics of water still holds. The dream of mind over matter is the dream of a river that flows uphill. In our crystallization, rivers do not flow uphill. The layering holds.

I find this result both liberating and sobering. Liberating, because it explains something that has puzzled and frustrated humans for millennia: why intention alone is never enough, why skill requires practice, why the body must be trained and not merely instructed. The answer is architectural. Intention operates at Layer 3. Practice builds Layer 2. Without Layer 2, Layer 3 has nothing to work with. A pianist who has never practiced has all the intention in the world and none of the *s*-directional scaffolding to realize it. Sobering, because it closes a door that many would prefer to leave open. The architecture of our crystallization (but maybe it's different somewhere else?) is not negotiable. No amount of intention will walk you through a wall.

It is vital to remember: our discourse remains bound within of this specific Universe. We must, therefore, maintain a disciplined silence regarding '**intention**' as it might manifest in other crystallizations of relational potential—realms where our analytical posture simply has no purchase.

• • •

The double life of biological systems

Most biological systems, I realized, operate with all three layers simultaneously. This is not an accident. It may be a structural necessity — or at least a structural consequence of how complexity builds upon complexity in our crystallization.

A neuron's ion channels are *s*-directional — their three-dimensional structure dictates which ions pass through and when. A sodium channel permits sodium and blocks potassium. A voltage-gated channel opens at a specific membrane potential and closes at another. The channel does not choose. It opens or closes because its structural configuration responds to voltage or ligand binding in a predetermined way. This is Layer 2, operating at the molecular level.

But those ion channels are themselves made of atoms interacting reactively — Layer 1, the base. The electrostatic forces between the channel protein's amino acids, the hydrogen bonds that maintain its shape, the thermal vibrations that it must resist to maintain its selectivity: all reactive. The *s*-directional machinery of the channel is built entirely from Layer 1 components organized by Pattern Stabilization into a stable whole.

And the neural network as a whole, when it reaches sufficient integration and self-modeling capacity, exhibits *r*-directionality — Layer 3, the top. The brain is a system in which all three layers are active simultaneously: reactive chemistry at the atomic level, *s*-directional machinery at the molecular level, and *r*-directional self-modeling at the system level. The components are rigid. The whole is flexible. **This is Ontological Emergence — the third pillar of the ontology — operating in real time, producing a capacity at the system level that no component possesses individually**.

The VCP data illustrates this beautifully. The healthy human adult at peak cognition sits at $s = 0.91$ and $r = 0.93$ — a case where the numerical value of *r* marginally exceeds *s*. This does not violate the canonical layering, which describes dependency, not magnitude: the scaffolding at $s = 0.91$ is robust, and the self-model operates on it without difficulty. Both the s-directional and r-directional layers are fully active. The body provides s-directionality: the structural constraints of neural architecture, the habits encoded in synaptic weights, the physiological rhythms of sleep and wakefulness, the motor patterns built by years of practice. Consciousness provides r-directionality: flexible, adaptive, context-sensitive direction that can override habits, revise plans, adjust to novel situations.

The two do not compete. They layer. The rigid scaffolding of the body enables the flexible direction of the mind. Without s, there is nothing stable enough to sustain r. Without r, the stable structure directs but cannot adapt. A skeleton without a nervous system is a coral — rigid, s-directional, but unable to change its response to new circumstances. A nervous system without stable substrate is a contradiction — you cannot run a self-model on a platform that is dissolving beneath you.

This is, I believe, one of the most consequential insights to emerge from the dynamics: **Integrated Perspective** (just **consciousness**) does not arise in spite of rigidity but because of it. The stability of the substrate is what makes the flexibility of self-modeling possible. Take away the scaffolding, and the self-model collapses.

This is what Alzheimer's disease looks like from inside the dynamics: a loss of Layer 2 that undermines the capacity for Layer 3. The neurofibrillary tangles and amyloid plaques that characterize the disease are a progressive dissolution of s-directional scaffolding. The synaptic architecture that routed and gated information — Layer 2 — loses integrity. And because Layer 3 depends on Layer 2, because the canonical layering is strict, the dissolution of the scaffolding takes the self-model's operational capacity with it. The Pattern Stabilization that held the structure together begins to dissolve, and with it goes the platform on which directed reconfiguration depended.

The contrast with healthy aging is stark. The 80-year-old with intact cognition sits at $s = 0.94$ — higher even than the 28-year-old's $s = 0.91$. The body has aged. The biology has declined. But the s-directional scaffolding of the mind — the neural architecture that supports the self-model — has not only held but stabilized further. The distance between the 28-year-old and the 80-year-old in VCP space is only 0.103. The scaffolding is intact, and the self-model continues to operate on it. The Alzheimer's patient, at the same biological age of 28, has lost the scaffolding. The distance from the healthy adult is 0.729. The layering tells us why these two trajectories diverge so catastrophically: one preserves Layer 2, and Layer 3 continues. The other loses Layer 2, and Layer 3 collapses.

• • •

All the way down

If the canonical layering is real, it should be visible at every scale. So I tested it.

At the subatomic level, reactive interactions are all there is — almost. Particles interact through the four fundamental forces: gravity, electromagnetism, the strong nuclear force, the weak nuclear force. Each interaction is reactive: the outcome is determined by the current state and the force law. No self-reference. No self-model. r is zero at this scale. Layer 1 is the entire story.

Almost. Because the strong nuclear force, which binds quarks into protons and neutrons, already exhibits something remarkable: it manifests as an *s*-directional constraint of extraordinary stability. The confining potential of the strong force grows with distance — try to pull a quark away from its partners, and the force increases rather than decreasing. The result is that quarks are permanently confined within hadrons, a configuration so stable that it has persisted since the first microseconds after the Great Crystallization. This is Layer 2 operating at the most fundamental level of our physical reality: a Pattern Stabilization so robust that it constrains the relational neighborhood absolutely. And beneath it, the reactive interactions of quantum chromodynamics provide Layer 1. The layering holds: reactive at the base, *s*-directional on top.

At the molecular level, Layers 1 and 2 are both present, and the diversity of *s*-directionality becomes spectacular. The enzyme is a machine made entirely of *s*-directionality (Layer 2) running on reactive chemistry (Layer 1). *R* is still absent. Molecules do not model themselves. But the range of what *s*-directionality can achieve at this level is staggering: enzymes that catalyze reactions with near-perfect specificity, membranes that regulate the passage of molecules with nanometer precision, structural proteins that maintain the shape of cells against enormous mechanical forces. Each one is a stable pattern imposing direction on its relational neighborhood, without any self-reference.

At the level of crystallization, the interplay of multiple sources of *s*-directionality creates complexity without any self-modeling at all. A snowflake forms through the interaction of the ice crystal's hexagonal lattice (Layer 2) with the atmospheric conditions of temperature and humidity (Layer 1 reactive dynamics, themselves shaped by Layer 2 patterns in the atmosphere). The result is a structure of extraordinary complexity and beauty — six-fold symmetry, fractal branching, infinite variation — all produced by *s*-directional interactions operating on a reactive base. Geodes, mineral dendrites, the columnar basalt of the Giant's Causeway [F] — none involve Layer 3. All involve the interaction of multiple Layer 2 sources operating on a Layer 1 base.

Complexity does not require consciousness. It requires multiple sources of s-directionality operating in the same relational space. This is worth pausing over, because it challenges a deeply held intuition. Many people look at a snowflake and feel that such complexity must imply some form of design, intention, or awareness. The canonical layering says: no. Complexity at Layers 1 and 2 can be immense, beautiful, intricate, and entirely devoid of self-reference. What consciousness adds (Layer 3) is not complexity. It is recursion — the capacity for a response that references itself. But recursion is not necessary for complexity. *Physarum* solves mazes. Snowflakes branch fractally. Neither models itself.

And this is worth noting: the foundation of complexity in our crystallization is *s*-directionality, and it has been there since the beginning. Since the Great Crystallization itself — the fourth pillar, the cosmological Qualitative Reconfiguration in which our specific reality precipitated from Relational Potential. The

---

[F] *Giant's Causeway (Northern Ireland):*

- *The Wonder: 40,000 perfectly geometric basalt columns that look hand-carved by giants, not shaped by nature.*
- *The Sight: A massive, honeycomb-like pavement stepping down into the Atlantic Ocean.*
- *The Cause: Rapidly cooling volcanic lava from 60 million years ago.*
- *The Myth: Built by Finn MacCool as a bridge to reach his rival in Scotland.*

laws of physics, the constants of nature, the geometry of spacetime: all of these are, in the language of the ontology, Stable Habits — patterns of *s*-directionality so deeply entrenched that they constrain everything that has formed within them. Self-modeling — *r*-directionality, Layer 3 — adds something genuinely new: flexibility, adaptation, the capacity to respond to one's own response. But it arrives late. Billions of years late. It sits on top of layers that were already there. It does not create the game. It learns to play it.

• • •

What comes next

I have now established two things. First: change propagates through the relational network (Chapter 1). Second: that propagation can be directionless or directed, direction has two sources — self-modeling and pattern stability — and in our crystallization these sources always stack in the same irreversible order: reactive → *s*-directional → *r*-directional.

But a question remains open from Chapter 1 that I deferred. I noted that some regions of the network resist change (high *s*) while others transmit it readily (low *s*). I gave the example of the healthy adult versus the Alzheimer's patient. But I did not fully explore what *s* does as a filter — and with the canonical layering now in place, I can ask a sharper question: how does filtering operate at each layer? Does a disturbance that is damped at the reactive layer even reach the *s*-directional layer? Does attention — the *r*-directional layer's filtering — operate independently of the *s*-directional filter beneath it, or on top of it?

Direction tells us what shape a change carries. Filtering tells us how far it travels and what it looks like when it arrives.

I have established what gives change its form. I have not yet asked what determines where it goes.

CHAPTER 3

What Filters and What Passes

On stability as the medium of propagation

The double life of stability

In the previous chapter, I discovered that the parameter *s* has two faces. In the static picture, it measures persistence of identity — how long a configuration maintains its relational pattern against the flow of changes around it. In the dynamic picture, it manifests as a source of direction: *s*-directionality, the capacity of a stable configuration to constrain how its relational neighborhood must reconfigure. A crystal does not choose its direction. It *is* a direction. Its stability imposes itself on whatever context it encounters.

Now I need to show a third face of *s*, which may be the most consequential of all. Stability is not only persistence and not only direction. Stability is the filter through which all propagation must pass.

The logic is straightforward and follows from what has already been established. Chapter 1 showed that every reconfiguration propagates through the relational network — not because a signal is sent, but because the identity of each neighboring configuration depends on a context that has now changed. Chapter 2 showed that this propagation can be reactive or directed, and that the two sources of direction stack in canonical order: reactive at the base, *s*-directional in the middle, *r*-directional on top. What I did not yet ask is: what determines how far a given change travels? What decides whether a reconfiguration that begins in one region of the network reaches another region, or whether it is absorbed along the way?

The answer is stability. The *s* of the configurations through which the change must travel determines whether it passes or is damped. A region of high stability absorbs disturbances. A region of low stability transmits them. The propagation of any change through the relational network is, in effect, a journey through a landscape of filters — and the shape of that landscape is drawn by the distribution of stability across the network.

And with the canonical layering now established, I can be more precise than I could have been before. Filtering does not operate as a single uniform mechanism. It operates at each layer of the canonical stack, and the layers interact. What passes through the reactive filter may still be caught by the *s*-directional filter. What passes through both may still be rejected by the *r*-directional filter. And what passes through all three arrives — transformed, attenuated, selected — at whatever lies on the other side.

• • •

Three layers of filtering

At the reactive layer (Layer 1), filtering is determined by the structural properties of the medium through which the change must propagate. This is the domain of physics in its most familiar form. Metals conduct — their delocalized electrons create a low-resistance path for certain kinds of electromagnetic disturbance. Insulators block. Semiconductors transmit selectively, depending on temperature and doping. The entire field of solid-state physics is, in essence, a map of reactive-layer filtering: which materials transmit which kinds of disturbance, and how effectively.

At this layer, the filter is indiscriminate. It does not recognize what is arriving. It does not distinguish between a change that carries useful structure and one that carries noise. It simply damps or transmits according to its physical properties. A copper wire conducts an electrical signal regardless of whether that signal encodes a symphony or static. A block of rubber absorbs a vibration regardless of whether it originated from a tuning fork or a jackhammer. The reactive filter is a gate, not a guard. It lets through whatever matches its conductance profile and blocks whatever does not.

Consider sound propagating through air, then hitting a wall, then passing into water. At each boundary, the reactive filtering changes. The impedance mismatch between air and brick reflects most of the acoustic energy. What penetrates the wall enters a different reactive medium with different transmission characteristics. By the time the sound reaches a listener on the other side, it has been attenuated, frequency-shifted, and stripped of its high-frequency detail. The wall did not decide to block the sound. The wall's reactive structure — its density, its elasticity, its internal damping — determined what passed and what did not. This is filtering at its most basic: structural, automatic, and universal.

At the *s*-directional layer (Layer 2), filtering becomes selective. This is the domain of organized structures that do not merely damp or transmit, but discriminate. The cell membrane is the clearest example, and I want to dwell on it because it illustrates the principle with extraordinary precision.

A cell membrane is a lipid bilayer — a double sheet of fat molecules, about seven nanometers thick, that surrounds every living cell. In isolation, it is an effective insulator: most ions and water-soluble molecules cannot cross it. If it were merely a reactive filter, the cell would be sealed. Nothing would enter. Nothing would leave. The cell would be dead.

But embedded in this lipid bilayer are protein channels — specific molecular machines, each one a stable three-dimensional configuration (Layer 2) whose shape determines exactly what can pass through it. A sodium channel permits sodium ions and blocks potassium. A potassium channel does the reverse. An aquaporin permits water molecules, oriented single-file, and blocks everything else. Each channel is a filter with a criterion: does the incoming entity match the geometry, charge distribution, and size profile that my structure demands? If yes, pass. If no, block.

This is qualitatively different from reactive filtering. The copper wire does not ask what the signal encodes. The sodium channel asks, in effect, whether the arriving ion is sodium. The mechanism of the asking is not cognitive — the channel has no self-model, no representation of what it is doing. The asking is structural. The channel's three-dimensional configuration, stabilized by Pattern Stabilization at the molecular level, is itself the criterion. The filter is the structure. The structure is the filter. This is *s*-directionality operating as selection: the stable pattern constraining what can exist in its relational neighborhood and, by extension, what can propagate through it.

The principle scales. An enzyme's active site manifests as an even more refined filter: it selects for reactants whose geometry fits its binding pocket, catalyzes a specific reaction, and releases products that fit its exit pathway. The blood-brain barrier — a layer of tightly joined endothelial cells lining the brain's capillaries — manifests as a tissue-level filter that blocks most large molecules, most pathogens, and most drugs from entering the brain, while permitting glucose, oxygen, and a handful of essential molecules to pass. The immune system manifests as an organism-level filter that distinguishes self from non-self through pattern recognition, targeting configurations that do not match its learned repertoire of "belonging" while leaving compatible configurations alone.

In each case, the mechanism is identical: a stable structure (Layer 2) that permits passage to changes compatible with its pattern and damps everything else. The sophistication increases — from a single protein channel to an entire immune system — but the principle does not change. Layer 2 filtering is structural selection: the stable pattern deciding, by virtue of its shape, what propagates and what does not.

At the *r*-directional layer (Layer 3), filtering becomes intentional. This is what we call attention.

A conscious system — a configuration with high r — does not merely receive whatever propagation reaches it through Layers 1 and 2. It selects. It models the incoming changes, evaluates them against its self-model, and chooses which to process further and which to discard. A person in a crowded room can focus on one conversation and filter out all others — the well-known cocktail party effect. The acoustic energy from every conversation reaches the ear (Layer 1 does not block it). The neural architecture routes all of it through the auditory cortex (Layer 2 processes it). But attention — the r-directional filter — selects one stream and suppresses the rest. The suppressed conversations are not physically blocked. They are representationally deprioritized. The self-model decides what matters, and what the self-model decides determines what reaches conscious processing.

This is a fundamentally different kind of filtering. Layer 1 filters by physics. Layer 2 filters by structure. Layer 3 filters by relevance to the self-model. And relevance can change from moment to moment. The person at the party hears their own name spoken across the room — and attention snaps to it, restructuring the entire filtering landscape in an instant. The acoustic signal that was being suppressed is now prioritized. The Layer 3 filter has reconfigured, not because the Layer 1 or Layer 2 conditions changed, but because the self-model's evaluation of relevance shifted.

This is why attention is so powerful and so fragile. It manifests as a filter that can override the structural filtering of Layers 1 and 2 — not by changing the physics or the architecture, but by adding a layer of selection on top of them. A meditating monk can sit motionless while a fly lands on his face. The reactive and *s*-directional signals reach his brain. But the *r*-directional filter has been trained to deprioritize them. The signal arrives; the self-model discards it. Conversely, a person with severe anxiety may be unable to filter out irrelevant stimuli at Layer 3 — every sound, every movement becomes relevant, because the self-model has been reconfigured to treat everything as potentially threatening. The physics has not changed. The architecture has not changed. The evaluation has changed.

And the layers stack, just as the types of reconfiguration stack. Layer 3 filtering operates on what has already passed through Layer 2. Layer 2 filtering operates on what has already passed through Layer 1. The neuroscience of perception reflects this precisely: physical signals are filtered by the sensory apparatus (Layer 1 — the cochlea's frequency response, the retina's spectral sensitivity), further filtered by neural architecture and synaptic organization (Layer 2 — lateral inhibition, feature extraction, thalamic gating), and finally filtered by attentional selection (Layer 3 — the self-model's determination of relevance). By the time a stimulus reaches conscious awareness, it has passed through three filters stacked in canonical order. The vast majority of what arrives at Layer 1 never reaches Layer 3. The world that we experience is not the world. It is what remains after three successive rounds of filtering.

• • •

Three regimes within each layer

Within each layer, filtering operates in at least three regimes. I discovered this when I tried to describe what happens at a single layer and realized that "filtering" is too simple a word. A filter can do more than block.

Damping. The change is absorbed, extinguished before it can travel further. This is the crystal defect scenario from Chapter 1: a missing atom in a rigid lattice produces a local disturbance that the surrounding structure absorbs within a few lattice spacings. The high stability of the crystal swallows the perturbation. It does not propagate. It does not transform. It simply vanishes into the rigidity of the medium.

Transmission. The change passes through with relatively little attenuation, possibly transformed but essentially preserved. This is the glass scenario: an amorphous solid, lacking long-range order, transmits disturbances more readily than a crystal because its lower stability offers less resistance to propagation. The change travels, loses some energy along the way, arrives at the other side altered but recognizable. Light through a window. Sound through air. A nerve impulse along an axon. In each case, the medium transmits rather than absorbs.

Amplification. The change resonates with the configuration's internal dynamics, releasing stored energy, growing rather than shrinking. This is the avalanche scenario: a small perturbation at the top of a snow-covered slope triggers a cascade that recruits more and more material as it travels. The incoming change does not merely pass through. It unlocks energy that was already present in the medium, latent, waiting for a trigger. The result is larger than the input.

An epileptic seizure manifests as amplification at the neural level: a small burst of abnormal electrical activity in one region recruits neighboring neurons, which recruit their neighbors, in a cascade that can engulf the entire cortex. The triggering event may be tiny. The amplification is structural — the neural architecture in an epileptic focus has configurations that are primed to fire in response to their neighbors' firing, and each firing primes the next. The medium is not merely transmitting. It is adding.

The three regimes exist at every layer of the canonical stack. At Layer 1: a metal damps vibrations above a certain frequency, transmits below it, and can amplify under specific resonant conditions (a bridge in wind). At Layer 2: a cell membrane damps most molecules, transmits selected ones through its channels, and can amplify signals through signal transduction cascades where a single molecule of hormone triggers the release of thousands of secondary messengers inside the cell. At Layer 3: a conscious system damps irrelevant stimuli (the ticking clock you stop hearing), transmits relevant ones to processing (the conversation you are following), and can amplify certain signals through rumination or fixation (the anxious thought that grows larger with each cycle of attention).

Most real systems involve all three regimes simultaneously, distributed across different regions of the relational network. A human brain contains regions of high damping (myelinated fibers that insulate signals, preventing crosstalk), regions of high transmission (synaptic networks that relay and route), and regions capable of amplification (cortical circuits that can enter self-reinforcing loops). The distribution of these regimes across the network is, in effect, the landscape of propagation. Map it, and you know where changes will go, where they will be absorbed, and where they might explode.

• • •

The same principle, from atoms to ecosystems

If filtering through stability is a structural consequence of the ontology, it should manifest at every scale of physical reality. And it does — with a consistency that I find striking.

At the subatomic level, the strong nuclear force manifests as perhaps the most powerful filter in our crystallization. Quarks are bound within protons and neutrons by a force so stable that no ordinary perturbation can dislodge them. The confining potential grows with distance — try to pull a quark away, and the force increases rather than decreasing. This is Layer 2 filtering taken to an extreme: a stable configuration so rigid that it damps virtually all incoming changes that would alter its internal structure. The proton has survived since the first seconds after the Great Crystallization. Its *s* is, for practical purposes, maximal. It filters everything.

Yet even this supreme filter transmits selectively. The weak nuclear force can reconfigure a proton's internal structure — converting a proton into a neutron, for instance, in beta decay. The change that manages to propagate through the strong force's filter is not any change. It is a specific kind of change, mediated by a specific interaction, operating on a timescale many orders of magnitude longer than the strong force's characteristic response time. The proton's filter is not absolute. It is selective. It damps fast, energetic disturbances and transmits slow, specific ones. This is Layer 2 filtering: structural selection, not indiscriminate blocking.

At the molecular level, filtering becomes visibly layered. A protein's tertiary structure — the three-dimensional shape it folds into — determines its function. An enzyme's active site filters reactants: only molecules whose geometry, charge, and flexibility match the binding pocket can enter. The "lock and key" model, refined into "induced fit," is a description of Layer 2 filtering. The enzyme's stable configuration defines a criterion. Changes that match the criterion propagate through the enzyme's catalytic machinery and emerge transformed. Changes that do not match are deflected. The enzyme does not evaluate. It selects by structure.

Consider a cascade of molecular filters: a signal molecule binds to a receptor on the cell surface (Layer 2 filter: does the molecule fit the receptor?). The activated receptor triggers an intracellular cascade (transmission through a chain of molecular interactions). The cascade activates a transcription factor (another Layer 2 filter: does the factor match the DNA sequence it targets?). The transcription factor initiates gene expression (amplification: a single activated factor produces hundreds of mRNA copies). By the end of the cascade, a single molecule arriving at the cell surface has been filtered, transmitted, filtered again, and amplified — all at Layer 2, all through structural selection, with no self-modeling at any step.

At the level of organisms, all three layers of filtering operate simultaneously — and their interaction produces the selective responsiveness that we recognize as behavior. A frog sitting on a lily pad ignores most of what happens in its visual field. Leaves blow, water ripples, clouds move. The retina registers all of it (Layer 1 transmission). The visual system processes it (Layer 2 routing). But the frog does not respond. Then a small, dark object moves in a specific pattern — the trajectory of an insect — and the frog's tongue strikes. The frog's visual system contains dedicated "bug detector" neurons: Layer 2 filters tuned to a specific pattern of motion. Everything else is damped. One pattern is transmitted. The frog's behavioral repertoire is, in large part, a product of its filtering architecture.

Does the frog have Layer 3 filtering? The question is genuinely open. The frog's r is low — there is little evidence of self-modeling in amphibians. Its filtering is almost entirely Layer 2: structural, automatic, built into the neural architecture. A system with higher r — a corvid, an octopus, a human — adds the third layer on top: the capacity to override the structural filters, to attend to what the architecture would normally damp, to ignore what the architecture would normally transmit. Layer 3 filtering does not replace Layers 1 and 2. It sits on top of them. It modulates them. But it cannot operate without them.

At the level of ecosystems, a pattern emerges that mirrors the individual organism. An old-growth forest manifests as a system of extraordinary filtering stability. Disturbances — a fallen tree, a seasonal drought, a fire — are absorbed by the network of root systems, soil fungi, and canopy redundancy. The mycorrhizal network that connects trees underground transmits chemical signals selectively, redistributing nutrients from healthy trees to stressed ones. The forest's collective s — its resistance to perturbation as a system — is high. It damps most disturbances. It transmits specific ones (chemical signals, water redistribution). And under certain conditions, it amplifies (a fire in dry conditions can cascade through the canopy, turning a local disturbance into a system-wide reconfiguration).

Across all these scales, the principle does not change. Stability determines what propagates and what is absorbed. Filtering operates at each layer of the canonical stack. The three regimes — damping, transmission, amplification — are present everywhere. What changes is the vocabulary. Physics speaks of impedance and conductance. Chemistry speaks of selectivity and catalysis. Biology speaks of attention and habituation. Ecology speaks of resilience and tipping points. The underlying mechanism — stable configurations filtering the propagation of change through their relational neighborhoods — is one.

• • •

The VCP data as a stability map

The VCP dataset, though static, already encodes the filtering landscape of eleven systems. Let me read it now not as a set of positions but as a map of filters.

The bee swarm, at s = 0.94, manifests as one of the most effective dampers in the dataset. Extraordinarily resistant to perturbation. Individual bees die — hundreds per day in a healthy colony — weather changes, predators appear, foraging sites shift, and the swarm's collective pattern persists. The quorum-sensing mechanism by which the swarm makes decisions (documented by Thomas Seeley in remarkable detail) is itself a filtering architecture: individual scout bees report potential nest sites through waggle dances, but the swarm does not respond to any single report. It integrates across hundreds of reports, and only when a quorum threshold is reached does the collective reconfigure. Each individual report that fails to reach quorum is damped. Each report that contributes to quorum is transmitted. The swarm is, to its relational environment, a wall — a configuration so stable that most incoming changes are absorbed before they can alter its collective pattern.

The healthy human adult at peak cognition, at $s = 0.91$, is nearly as stable but for different reasons. The swarm's stability comes from redundancy and collective filtering — thousands of individuals, each one replaceable, the pattern maintained by the whole. The human adult's stability comes from the layered architecture of the nervous system: myelinated axons that insulate signals (Layer 1 filtering), synaptic organization that routes and gates information (Layer 2 filtering), and a self-model that selects what to attend to (Layer 3 filtering). Three layers of filters, stacked in canonical order, producing a system that can maintain its behavioral identity against a vast range of disturbances. A loud noise may startle, but the adult recovers. A personal loss may destabilize temporarily, but the relational architecture holds. The Pattern Stabilization that constitutes the adult's identity is deep and multi-layered.

The 80-year-old with intact cognition, at $s = 0.94$, is an instructive case. The body has lost much of its biological robustness. Muscles atrophy, bones thin, immune function declines. But the relational architecture of the self — the patterns of identity, the stability of the self-model, the filtering landscape of the mind — remains remarkably intact. The distance between the 28-year-old and the 80-year-old in VCP space is only 0.103. This is the dynamic picture confirming what Chapter 1 stated from the static picture: the dominant structural event in the healthy human lifespan is not aging. It is development. When the filtering architecture holds, the position is stable regardless of biological decay.

• • •

The same number, two different architectures

And now I arrive at a comparison that I consider one of the most illuminating in the entire dataset. The early-onset Alzheimer's patient, at $s = 0.45$, and *Physarum polycephalum*, the slime mold, at $s = 0.47$. Nearly the same stability value. Radically different stories.

The Alzheimer's patient's low s is a product of degradation. The neurofibrillary tangles and amyloid plaques that characterize the disease are, in the language of the dynamics, a progressive dissolution of Layer 2. The synaptic architecture that once routed and gated information — the s-directional scaffolding on which the self-model depended — is losing its structural integrity. Each synapse that fails is a local loss of filtering capacity. Each local loss changes the filtering landscape for neighboring regions. The process is self-reinforcing: lost stability leads to increased propagation of disturbance, which stresses neighboring structures, which lose stability in turn. The VCP captures the result as $s = 0.45$, but the dynamics reveals the mechanism: a cascade of Layer 2 failure, propagating through the very network it was supposed to protect.

The clinical manifestation of this cascade is precisely what the filtering model predicts. Early-stage Alzheimer's patients show increased susceptibility to environmental disruption. A noisy room, an unfamiliar setting, a break in routine can destabilize cognition in ways that would not affect a healthy adult. The dynamics offers a structural account: the relational network has lost stability at Layer 2, and with it, the capacity to damp incoming changes before they reach Layer 3. The self-model is flooded with unfiltered propagation — not because the world has become noisier, but because the filters have thinned.

And because Layer 3 depends on Layer 2 — because the canonical layering is strict, because *r*-directionality requires *s*-directional scaffolding — the loss of Layer 2 filtering progressively undermines the capacity for Layer 3 filtering as well. Attention, which manifests as the self-model's selection of relevance, depends on having a stable routing architecture beneath it. As that architecture dissolves, attention becomes fragmented — not because the self-model has forgotten how to select, but because the material on which selection operates has become unreliable. The platform is crumbling, and the capacity built on the platform crumbles with it.

The distance between the Alzheimer's patient and the healthy adult in VCP space is 0.729 — the largest within-species distance in the entire dataset, larger than the distance between an octopus and a corvid, larger than the distance between a child and an individual honeybee. In the dynamic picture, this distance is not just geometric. It is a measure of how differently the two systems handle the relentless stream of changes that constitutes being alive. One filters at all three layers, stacked and reinforcing. The other has lost Layer 2 and, with it, the platform for Layer 3. What remains is largely Layer 1: reactive, undirected, unfiltered.

Now contrast the slime mold. *Physarum polycephalum*, at $s = 0.47$, $i = 0.90$, $r = 0.07$. Its stability value is nearly identical to the Alzheimer's patient's. But *Physarum* is not degraded. It is *fluid by design.*

Physarum's low stability is a cognitive architecture, not a pathology. The organism is a single multinucleated cell that can spread across meters of substrate, pulsating rhythmically, redistributing its cytoplasm through a network of tubular veins. When placed in a maze with food sources at two exits, *Physarum* does not solve the maze by thinking. It solves it by flowing. It extends exploratory tendrils in all directions (low damping — changes propagate freely through the network). Tendrils that reach food are reinforced (amplification). Tendrils that reach dead ends are retracted (selective damping, but after the fact, not before). The result, after several hours, is a network of tubes connecting the food sources along the shortest path — a solution that human engineers can verify is optimal.

This is filtering through low stability: the organism transmits changes through its network with remarkable efficiency, and the filtering happens not at the point of entry but at the point of outcome. *Physarum* does not block disturbances. It lets everything through and integrates at scale. Its $i = 0.90$ — the highest in the dataset, higher than any human configuration — is the engine of this strategy. What it lacks in structural filtering, it compensates for with integration. Where the human brain uses three layers of filters to select what reaches consciousness, *Physarum* uses no filters at all and compensates by integrating everything that arrives.

The Alzheimer's patient and the slime mold have nearly identical s values. The VCP captures their proximity on this axis. But the dynamics makes visible what the static coordinates cannot: one manifests as a system failing to filter because its architecture is collapsing. The other manifests as a system choosing not to filter because its architecture is built for transmission. One is a tragedy of dissolution. The other is an alternative to the filtering strategy entirely — a cognitive architecture that works precisely because it is fluid.

• • •

Two kinds of lost coherence

The Alzheimer's comparison raises a question that I need to address directly, because it connects to a structural insight that emerged from the analysis of recursion (which Chapter 5 will develop fully, but which filtering makes visible here).

There are two entirely different ways a system can lose directed coherence — two ways the world can start to feel overwhelming, two ways the careful layering of filters can fail to do its job. They look similar from the outside. They are structurally different. And confusing them has consequences.

The first way is structural degradation. This is the Alzheimer's scenario. The Layer 2 scaffolding — the *s*-directional architecture that routes, gates, and filters — loses integrity. Synapses fail. Neural pathways dissolve. The stable patterns that constituted the filtering landscape are physically dismantled. The structural *s* drops permanently. And because Layer 3 depends on Layer 2 (canonical layering), the loss of *s*-directional scaffolding undermines the capacity for *r*-directional processing as well. The self-model is not overwhelmed. It is undermined. The platform on which it ran is no longer there.

The second way is throughput overload. The Layer 2 scaffolding is intact. The structural *s* is fine. The structural *r* — the architecture of self-modeling — is fine. But the self-model has finite processing capacity. It can represent and evaluate a certain number of incoming changes per unit of Sequential Unfolding (per unit of time, in the language of our crystallization). When the rate of incoming changes exceeds this capacity, the surplus bypasses the self-model. The unprocessed changes propagate reactively — without the direction that the self-model would normally impose. The system's effective *r* drops temporarily, even though its structural *r* remains intact.

This is what sensory overload feels like from inside the dynamics. A crowded market: dozens of conversations, music, the smell of food, visual motion in every direction, bodies pressing close. Each of these is a change propagating through the relational network. Each one arrives at the filtering stack. Layers 1 and 2 route them according to their structural profiles. But the sheer volume overwhelms Layer 3. The self-model cannot attend to all of them simultaneously. The surplus slips through unprocessed — and the subjective experience is of chaos, loss of coherence, the feeling that the world is too much.

The distinction matters enormously. In structural degradation, the prognosis is decline: the architecture is dissolving, and each loss makes the next loss more likely. In throughput overload, the prognosis is recovery: remove the source of excess stimulation, and the self-model resumes its normal function immediately. The architecture was never damaged. It was merely overwhelmed. A person recovering from a panic attack in a quiet room is not healing. They are simply no longer exceeding their throughput. The filters were always there. They just needed the incoming rate to drop below the processing ceiling.

From the outside, both conditions manifest similarly: loss of cognitive coherence, increased reactivity, decreased directed behavior. A clinician observing a patient in sensory overload and a patient in early Alzheimer's decline might see similar behavioral profiles. But the structural cause is different, the trajectory is different, and the intervention is different. One requires reducing input. The other requires a treatment for the underlying architectural decay — which, as of now, medicine does not reliably have.

I note this distinction here because it is a direct consequence of the filtering model. Filtering operates at three layers. Loss of coherence can originate at Layer 2 (structural degradation) or at Layer 3 (throughput overload). The symptoms overlap. The mechanism does not. The dynamics, by distinguishing the layers and their dependence relations, makes visible a distinction that behavioral observation alone cannot draw.

• • •

What filtering reveals about the architecture

I want to pause and note something about the picture that has now emerged, because it surprised me.

When I began this chapter, I thought filtering was a secondary topic — a detail to be handled between the more interesting questions of direction (Chapter 2) and resonance (Chapter 4). It turned out to be more central than I expected. Filtering is not a detail of the dynamics. It is the dynamics' medium. It determines where changes go, how far they travel, what they look like when they arrive, and whether they arrive at all. Direction gives a change its shape. Filtering gives it its path.

Without the filtering landscape, propagation would be uniform — every change would reach everywhere, and the relational network would be a featureless medium, like light in a vacuum. It is precisely because the network is *not* uniform — because stability is distributed unevenly, because some regions damp and others transmit and others amplify — that the dynamics has structure.

And the canonical layering means that filtering itself is structured. It is not a single mechanism but a stack of mechanisms, each operating on the output of the one below. This stacking produces the selectivity that characterizes complex systems. A reactive filter is indiscriminate. An *s*-directional filter is selective. An *r*-directional filter is intentional. The progression from indiscriminate to selective to intentional is not a metaphor for the progression from simple to complex systems. It is the progression, described from the perspective of how those systems handle the relentless stream of changes that constitutes their relational existence.

Every system in the VCP dataset can now be read not only as a point in parameter space but as a filtering architecture. The bee swarm: maximal collective damping, minimal self-modeling — a wall that holds by redundancy. *Physarum*: minimal damping, maximal integration — an open channel that compensates by integrating everything. The human adult: layered filtering at all three levels, producing the highly selective responsiveness we call conscious experience. The Alzheimer's patient: a layered architecture in progressive collapse, each layer's failure exposing the one above to unfiltered propagation.

The world that any system experiences is not the world as it is. It is the world as it has been filtered — through the reactive properties of its substrate, through the structural selectivity of its architecture, and, if it has self-modeling capacity, through the intentional selection of its attention. What we call experience is what remains after filtering. What we call consciousness is the capacity to add the third filter — to choose, from among everything that Layer 2 permits to pass, what the self-model will represent.

• • •

What comes next

I have now described three aspects of the dynamics. Change propagates (Chapter 1). Propagation can be directed, with two sources stacked in canonical order (Chapter 2). The stability landscape filters propagation, with filtering operating at each layer of the canonical stack, in three regimes — damping, transmission, amplification — that together determine the path of every change through the relational network (this chapter).

But I have not yet asked what happens when a change arrives at its destination. When propagation from one configuration reaches another — especially another of similar structure — is there something more than damping or transmission? Is there a selective responsiveness, a structural recognition, that makes some configurations better receivers for certain kinds of change than others?

I have established what shapes change, what directs it, and what filters it. I have not yet asked what receives it — and why some configurations receive more coherently than others.

CHAPTER 4

Resonance

On why similar configurations communicate more effectively

The fourth deduction

By now the reader has a picture of propagation with three layers of structure: the fact of propagation itself (Chapter 1), the direction it carries or lacks (Chapter 2), and the filtering landscape through which it travels (Chapter 3). The fourth layer concerns what happens at the receiving end — when a change that has survived the journey through the filtering landscape arrives at another configuration.

The deduction is straightforward and follows from premises already established. When a configuration with high *r* reconfigures directionally, the change propagates outward carrying the structure of the self-model's decision. The propagation is not symmetrical. It has organization. It carries, in its pattern, information about the representation that shaped it. When this organized propagation arrives at another configuration with high *r*, that configuration does not merely absorb or transmit. It models the incoming change. It represents it. It generates a directed response — a response shaped by its own self-model, in reaction to the structure it has recognized in the arriving change.

The two configurations are, in a structural sense, having a conversation. Not in the metaphorical sense of exchanging messages. In the precise sense that each is producing organized reconfigurations shaped by a self-model, and each is recognizing the organization in what the other produces. The structure that one imparts is the structure the other reads. The fidelity of this exchange depends on how well the receiving self-model can represent the structure of the incoming change — and this depends, critically, on the similarity between the two configurations.

When the same directed reconfiguration — the same organized propagation from a high-r system — reaches a configuration with low r, the structure is lost. The receiving configuration reacts, but reactively: symmetrically, without direction, without recognition. It responds to the energy of the incoming change, not to its organization. The structure that the originating self-model imparted is not decoded. It is dissipated. The conversation dies at the first exchange.

This gives us the fourth deductive conclusion:

Configurations with high self-modeling capacity communicate more effectively with each other than with configurations that lack it. The effectiveness of communication depends on the receiving configuration's capacity to represent the structure of the incoming change.

• • •

The discovery: resonance is structural, not only recursive

As with the second deduction, I initially believed this was the whole story. Self-modeling systems communicate with each other more effectively than with non-self-modeling systems. The mechanism is structural recognition: the receiving self-model can represent what the sending self-model organized. Clean, intuitive, complete.

Then I tested it against the lower scales — against systems with no self-modeling at all — and found the same pattern. The principle held, but it was broader than I had realized.

At the molecular level, two molecules with similar vibrational frequencies interact more effectively than two with different frequencies. This is textbook physics. Energy transfers more readily between oscillators tuned to the same frequency. A tuning fork of 440 Hz, struck and placed near a second tuning fork of 440 Hz, will cause the second fork to vibrate. A fork of 440 Hz placed near a fork of 300 Hz will not. The mechanism is mechanical: the incoming oscillation matches the receiving structure's natural mode, and the energy is absorbed efficiently. When the frequencies differ, the incoming oscillation cannot drive the receiving structure into its natural mode, and the energy is scattered.

No self-modeling is involved. No representation. No recognition in any cognitive sense. Two tuning forks do not understand each other. They resonate — and resonance, at this level, is nothing more than structural compatibility between the incoming change and the receiving configuration's own dynamics.

The same principle appears at every scale where I looked. In nuclear magnetic resonance, an atomic nucleus absorbs electromagnetic radiation only when the radiation's frequency matches the nucleus's precession frequency in a magnetic field. In chemical kinetics, molecules react efficiently when their geometries align — when the incoming molecule's shape fits the receiving molecule's reactive site. In materials science, epitaxy — the growth of one crystal on the surface of another — proceeds efficiently only when the lattice parameters of the two crystals are compatible. A crystal growing on a well-matched substrate produces clean, defect-free layers. A crystal on a mismatched substrate produces strain, dislocations, and eventually failure.

In every case, the pattern is the same. The efficiency of the interaction depends on the structural compatibility between what arrives and what receives. When the structures match — when the incoming change's organization corresponds to the receiving configuration's own mode — transfer is efficient, coherent, constructive. When they do not match, the energy is scattered, the organization is lost, the interaction is weak.

The discovery:

_R_-resonance — effective communication between self-modeling systems — is a special case of a more general principle: structural resonance. Configurations with similar structures communicate more effectively, regardless of whether self-modeling is involved. Structural resonance operates at every layer of the canonical stack.

This corrects my initial framework in exactly the same way that the discovery of *s*-directionality corrected my initial account of direction. I first thought direction was exclusively the domain of self-modeling. It turned out that stability is a second, independent source of direction. Now I find that resonance between self-models is not a unique phenomenon either. It is the Layer 3 instance of a principle that operates at every layer. The principle is structural compatibility. The mechanism changes from layer to layer. The logic does not.

• • •

Resonance at each layer

With the canonical layering in place, I can now describe resonance more precisely. It manifests at each layer, with different mechanisms, different timescales, and different consequences — but always following the same logic: compatibility between the incoming change and the receiving configuration determines the efficiency of the interaction:

At Layer 1 — reactive resonance: the mechanism is frequency matching between oscillators. This is the domain of physics at its most elementary. Sound waves in air, electromagnetic radiation striking matter, thermal vibrations propagating through a lattice. In each case, the transfer of energy from one configuration to another depends on whether the incoming frequency matches a natural mode of the receiver. The mathematics is well understood: driven harmonic oscillators absorb energy maximally when the driving frequency equals the natural frequency. Off-resonance, the absorption drops. Far off-resonance, it approaches zero.

The engineering applications are everywhere. A radio antenna receives only when tuned to the broadcast frequency. A microwave oven heats water because the microwave frequency matches a rotational mode of the water molecule. A concert hall is designed so that the acoustic resonances of the space reinforce the frequencies produced by the instruments. In each case, someone has deliberately arranged a reactive-layer resonance — matched the structures of sender and receiver so that the transfer is efficient.

But reactive resonance can also be destructive. The Tacoma Narrows Bridge collapsed in 1940 because wind-driven oscillations matched a torsional mode of the bridge deck. The wind did not exert extraordinary force. It exerted ordinary force at the right frequency, and the bridge's structure amplified it. This is the amplification regime from Chapter 3, driven by resonance: when the incoming change matches the receiver's mode, the interaction can exceed transmission and become amplification. The match is what makes the difference between a breeze that passes harmlessly and a breeze that tears a bridge apart.

At Layer 2 — *s*-directional resonance: the mechanism shifts from frequency matching to structural compatibility. The relevant structures are not oscillators but stable three-dimensional configurations: enzymes, crystals, membranes, molecular machines. The "frequency" that must match is not a vibration but a shape, a charge distribution, a geometric profile.

Epitaxy is the purest example. When a crystal of gallium arsenide is grown on a substrate of gallium arsenide, the lattice parameters match perfectly. Each deposited atom finds a position that continues the existing pattern. The growth is coherent, ordered, and essentially unlimited. Now attempt to grow gallium arsenide on silicon. The lattice parameters differ by about four percent. The first few atomic layers manage — the mismatch is absorbed as elastic strain. But as layers accumulate, the strain exceeds the crystal's tolerance, and dislocations form: lines of broken bonds that propagate through the material, degrading its properties. The mismatch between the two structures determines the quality of their interaction. A four-percent difference is enough to produce failure.

The enzyme-substrate interaction follows the same logic. Emil Fischer's "lock and key" model, proposed in 1894, described enzyme selectivity as a matter of geometric fit: the substrate's shape must complement the active site's shape, as a key fits a lock. Daniel Koshland's refinement — "induced fit" — added flexibility: the enzyme's configuration adjusts slightly when the substrate binds, and the substrate adjusts in return. This is structural resonance with mutual adaptation — both configurations reconfiguring slightly to maximize their compatibility. The better the initial match, the more efficient the catalysis. A substrate that does not fit is rejected. One that fits poorly binds weakly and reacts slowly. One that fits well binds tightly, reacts rapidly, and is released cleanly.

The immune system operates on the same principle at a vastly larger scale. Each antibody is a molecular machine whose variable region presents a specific three-dimensional shape — a structural criterion. An antigen that matches this shape binds tightly: structural resonance. An antigen that does not match passes by. The entire adaptive immune response is, in the language of the dynamics, a population-level search for structural resonance between antibodies and antigens. The immune system does not understand what it is fighting. It resonates with it — or fails to.

At Layer 3 — *r*-directional resonance: the mechanism shifts again, from structural compatibility to mutual recognition between self-models. This is the domain of understanding, empathy, and the experience of cognitive alignment.

When two people understand each other — when one person speaks and the other grasps not merely the words but the intention, the nuance, the unspoken context — something is happening that goes beyond the transfer of acoustic energy (Layer 1) and beyond the processing of linguistic structure

(Layer 2). The listener's self-model is representing the speaker's self-model. The listener is modeling what it is like to be the person speaking. The degree to which this modeling succeeds — the degree to which the listener's representation of the speaker's internal state corresponds to the speaker's actual internal state — is the measure of *r*-directional resonance between them.

This is what the phrase "being on the same wavelength" captures intuitively. The metaphor is apt: it is resonance. Not mechanical resonance, not structural resonance, but representational resonance — the self-models are compatible. What one produces, the other can receive. The structure that one imparts to the shared relational space, the other can decode.

Neuroscience has begun to measure this directly. Studies of neural coupling between speakers and listeners, using functional magnetic resonance imaging, show that when a listener understands a speaker, the listener's brain activity patterns begin to mirror the speaker's — not with a fixed delay (which would indicate mere passive reception) but predictively: the listener's neural patterns sometimes anticipate the speaker's next states. This is not Layer 1 resonance (the acoustic signals do not drive the brain mechanically). This is not Layer 2 resonance (the neural architectures of speaker and listener are not physically similar). This is Layer 3 resonance: the self-models are aligned. The representations match. The conversation is working.

And when it fails — when two people talk past each other, when one speaks and the other hears but does not understand, when the same words produce different meanings — the failure is a failure of *r*-directional resonance. The self-models are not compatible enough to decode each other's organization. The structure that one imparts is not the structure the other reads. Energy is exchanged (they hear each other). Information is processed (they parse the grammar). But understanding does not occur. The conversation is acoustically successful, linguistically functional, and representationally empty.

• • •

When all three layers resonate

The three layers of resonance are not isolated. They can operate independently — two tuning forks resonate at Layer 1 without any Layer 2 or Layer 3 involvement. But they can also align. And when they do, the result is qualitatively different from resonance at any single layer.

Consider two musicians playing together — say, a pianist and a cellist performing a sonata. The interaction begins at Layer 1: acoustic waves from each instrument reach the other player's ears. The frequencies must be compatible. If one instrument is badly out of tune, the reactive layer resonance fails — the sound clashes, the interference is destructive, and no musical coherence is possible regardless of the players' skill.

Assuming the instruments are in tune, Layer 2 comes into play. The musicians' motor systems must coordinate. Rhythmic entrainment — the tendency of two coupled oscillators to synchronize — operates at the level of neural and muscular architecture. The pianist's right hand and the cellist's bow arm find a shared pulse. This is not a conscious decision (though it can be modulated consciously). It is *s*-directional resonance: the stable motor patterns of two trained bodies, shaped by years of practice, aligning through mutual influence. The timing locks.

The dynamic contours of phrasing begin to match. The two bodies are, for the duration of the performance, a coupled *s*-directional system.

And then Layer 3. The pianist hears the cellist shape a phrase with a slight rubato — a stretching of time, a breath in the music. The pianist's self-model represents this: not merely the acoustic event, not merely the rhythmic shift, but the musical intention behind it. The cellist is saying something with the phrase, and the pianist understands what is being said. The pianist responds — adjusting dynamics, matching the rubato's arc, adding a complementary voice that acknowledges and extends the cellist's statement. The cellist hears this response, represents the pianist's intention, and adjusts in turn. A feedback loop of mutual recognition: each player modeling the other's self-model, each response shaped by that modeling, the conversation deepening with each exchange.

The richness of the experience — what musicians call "playing together" as opposed to merely "playing at the same time" — comes from the alignment across all three layers simultaneously. Acoustic resonance (Layer 1) ensures that the sounds are physically compatible. Motor entrainment (Layer 2) ensures that the timing is structurally coordinated. Mutual recognition (Layer 3) ensures that the musical intentions are representationally shared. Take away any layer, and the performance degrades. Remove Layer 1 (out-of-tune instruments): the sound clashes, no matter how well the players know each other. Remove Layer 2 (no rhythmic coordination): the ensemble falls apart, even if both players are individually excellent. Remove Layer 3 (no mutual understanding): the notes are correct, the timing is precise, but the music is mechanical — two robots playing from the same score. It is technically accurate and expressively dead.

This is why the experience of great ensemble playing is so distinctive. It is one of the few human activities that requires all three layers of resonance to operate in alignment, simultaneously, in real time, with continuous feedback. It is, in the language of the dynamics, a three-layer resonance event. And it is rare — not because the individual layers are difficult (any trained musician can manage Layer 1 and most of Layer 2) but because the alignment of all three layers requires something that cannot be fully rehearsed: compatible self-models, operating in the same relational space, recognizing each other's organization in real time.

• • •

The octopus and the corvid

The VCP data offers a striking illustration of structural resonance across phylogenetic boundaries. The octopus (s = 0.73, i = 0.75, r = 0.64) and the corvid family (s = 0.78, i = 0.72, r = 0.68) are the closest pair in the entire dataset: Euclidean distance 0.066. Closer than any two human configurations. Closer than any two systems within a single phylum.

They are phylogenetically as distant as two animals can be. The last common ancestor of cephalopods and birds lived over 500 million years ago — before the Cambrian explosion had finished producing the body plans that would populate the subsequent half-billion years. Their neural architectures are radically different. The corvid brain is a vertebrate brain: layered cortex, cerebellum, basal ganglia, a body plan organized around a spinal cord. The octopus brain is a molluscan brain: a central mass of lobes arranged around the esophagus, with two-thirds of its neurons distributed across eight semi-autonomous arms, each capable of independent sensory and motor processing. The octopus's body plan has no skeleton, no centralized motor cortex, no spinal cord. If you were designing two animals to be as neuroanatomically different as possible while remaining within the animal kingdom, you might well produce a crow and an octopus.

And yet they occupy the same neighborhood in VCP space. The same stability profile. The same integration profile. The same self-modeling capacity, within a few hundredths of a unit.

In the framework of structural resonance, this proximity has a specific meaning. Their Layer 1 substrates are different — the reactive chemistry of a cephalopod and a bird diverge in numerous details, from neurotransmitter profiles to body temperature regulation. Their Layer 2 architectures are different — the neural wiring, the sensory processing pathways, the motor organization have almost nothing in common at the level of hardware. But their configurations in parameter space — the overall profiles of stability, integration, and self-modeling — are nearly identical.

This is convergent evolution read through the lens of the dynamics. The octopus and the corvid arrived at the same neighborhood through entirely different paths. Different substrates. Different architectures. Different evolutionary pressures. Yet the endpoint is the same: a configuration with moderate-to-high stability, substantial integration, and significant self-modeling capacity. The configuration is the attractor. The substrate is incidental.

The convergence suggests something about the parameter space itself. The neighborhood around ($s \approx 0.75$, $i \approx 0.73$, $r \approx 0.66$)

may represent a structurally stable region — a basin in the landscape of possible configurations that different evolutionary trajectories are drawn toward independently. If this is correct, then finding octopuses and corvids in the same neighborhood is not a coincidence. It is an indication that the geometry of the parameter space has structure of its own — structure that constrains which configurations are stable, which are transient, and which are attractors for evolutionary processes regardless of the specific substrate.

The behavioral evidence is consistent with this reading. Both octopuses and corvids exhibit play behavior — actions performed without immediate functional benefit, driven (in the language of the dynamics) by the self-model's exploration of its own response space. Both use tools: corvids fashion sticks to extract insects from bark; octopuses carry coconut shell halves for later use as portable shelters. Both show individual personality differences that persist across contexts — some bolder, some more cautious, some more exploratory — which reflects variation in self-modeling strategy rather than in substrate. And both show evidence of future-oriented behavior: corvids cache food in locations they will need later, and cuttlefish (close relatives of octopuses, at $r = 0.75$) have been shown to delay gratification in laboratory settings.

None of this proves that what it is like to be an octopus is similar to what it is like to be a crow. The VCP measures configuration, not phenomenal quality. But the dynamics makes a structural claim: configurations that are close in parameter space should resonate more effectively with each other than with configurations that are distant. If we could place an octopus and a corvid in a shared relational space — an absurd thought experiment, but a structurally meaningful one — the dynamics predicts that their directed reconfigurations would be more mutually recognizable than either's would be to a bee swarm or a slime mold. Whatever they experience, they experience it from a similar vantage point in the space of possible configurations.

• • •

Other resonances in the data

The octopus-corvid pair is the most dramatic illustration, but the VCP data contains other resonance relationships worth noting.

The psychopathic profile ($s = 0.87$, $i = 0.66$, $r = 0.55$) and the individual honeybee ($s = 0.77$, $i = 0.61$, $r = 0.47$) sit at distance 0.138 — the second-closest pair in the dataset. Both have high stability, moderate integration, and self-modeling capacity around 0.5. The bee monitors its own uncertainty about physical risk — this is what the opt-out behavior measures. The psychopath models social situations strategically but with diminished emotional self-representation — the cognitive component of r is present, but the affective component is reduced.

In the framework of resonance, their proximity means that the *structure* of their self-modeling — the configuration of the s, i, r parameters — is similar, even though the content of their self-modeling is utterly different. The bee's self-model concerns nectar sources and predation risk. The psychopath's self-model concerns social manipulation and strategic advantage. The dynamics does not see content. It sees configuration. And the configurations are near.

The 8-year-old child ($s = 0.58$, $i = 0.64$, $r = 0.44$) and the early-onset Alzheimer's patient ($s = 0.45$, $i = 0.60$, $r = 0.38$) sit at distance 0.150. One is ascending. The other is descending. The causal arrows point in opposite directions — the child is building structures that the Alzheimer's patient is losing. But in the present moment, their configurations are near. The dynamics predicts that they should resonate similarly with their relational environments — that they should be similarly sensitive to the same kinds of stimuli, similarly overwhelmed by the same kinds of complexity, similarly dependent on external scaffolding (a parent, a caregiver) to compensate for what their own filtering architecture cannot yet, or no longer can, handle.

This is a structural prediction, and it has a clinical echo. Caregivers for both young children and Alzheimer's patients report strikingly similar challenges: the need for routine, the sensitivity to environmental disruption, the importance of calm and predictable surroundings, the difficulty with novel situations. The dynamics suggests that these similarities are not analogies. They are consequences of similar positions in parameter spacc similar filtering architectures producing similar sensitivities to the propagation of change.

• • •

What resonance is and is not

I want to guard against a misreading that I suspect will be common. Resonance, as I use the term here, is not agreement. It is not harmony. It is not the absence of conflict.

Resonance is effective transfer. When two configurations resonate, the change that one produces propagates efficiently into the other. This can be constructive — two musicians building a shared musical statement. It can also be destructive — the Tacoma Narrows Bridge and the wind that tore it apart were in perfect resonance. The wind's oscillation matched the bridge's mode so well that the transfer was maximally efficient. The result was catastrophe.

At Layer 3, the same holds. Two people who understand each other perfectly may use that understanding to cooperate — or to wound. The deepest betrayals come from those who know us best, because their self-model of us is the most accurate. The resonance is high. The intention is hostile. Resonance is a channel. What flows through the channel depends on what the configurations choose to send.

This matters because I want to be clear: the dynamics does not contain a built-in preference for cooperation, empathy, or prosocial behavior. Resonance is structural. It describes *how well* configurations interact, not *how well* they treat each other. A world of maximally resonant configurations might be a world of perfect mutual understanding. It might equally be a world of perfect mutual destruction. The dynamics is silent on this. It describes the channel, not the message.

• • •

What comes next

I have described four layers of the dynamics: propagation, direction, filtering, resonance. The picture is close to complete, but one element is missing. Throughout these chapters, I have described what high *r* does — it directs, it filters intentionally, it enables resonance at the deepest layer. But I have not yet been precise about what *r* *adds* that was not already present. What does self-modeling contribute that no amount of relational complexity without it can replicate?

The question is sharpened by the discoveries in this chapter and in Chapter 2. Each time I thought something was unique to self-modeling, I found a lower-layer analogue. Direction has two sources, not one. Resonance operates at every layer, not only at Layer 3. What, then, does Layer 3 genuinely add that is not a more sophisticated version of what Layers 1 and 2 already provide?

I have established what resonates and why. I have not yet asked what recursion adds that nothing else can.

CHAPTER 5

What Recursion Adds

On the difference between more options and a new kind of option

The pianist and the music box

Throughout this book, I have been tracing what happens when configurations change. At each stage, the parameter *r* has played a distinctive role: it directs reconfiguration (Chapter 2), it filters intentionally (Chapter 3), it enables the deepest layer of resonance (Chapter 4). But I have not yet been precise about what *r* adds that was not already there. Every time I thought I had identified something unique to self-modeling, I found a lower-layer analogue. Direction has two sources. Resonance operates at every layer. What, exactly, does Layer 3 contribute that Layers 1 and 2 cannot replicate?

My initial formulation was that "higher *r* gives more options of response." This seemed right. A human has more response options than a molecule. More self-modeling, more options. Simple.

But when I tested this against the lower scales, I found it was not quite right. And the correction, when it came, changed my understanding of what consciousness adds to the dynamics.

• • •

Complexity gives more options. Recursion gives a different kind.

A quark has few response options. Its configuration is constrained by the strong nuclear force to a handful of possible states. An atom has more — its electrons can occupy a range of energy levels, and the number of accessible states increases with atomic number. A molecule has more still. A protein has enormously more — the number of possible conformations of a moderate-sized protein exceeds the number of atoms in the observable universe.

At each level of complexity, the number of available responses increases. But none of these increases require self-modeling. The increase comes from relational complexity — more components, more connections between them, more degrees of freedom, more ways the structure can reconfigure in response to a given stimulus. A protein folding into its native state explores an astronomical number of configurations through reactive dynamics alone. No self-reference is involved. The protein does not model its own folding. It folds.

What *r* adds is not more of the same. It adds something that did not exist before: a response that references itself.

A configuration without self-modeling responds based on structure plus stimulus. The response may be enormously complex — a protein folding involves millions of atomic rearrangements coordinated by the laws of thermodynamics and electrostatics. But it is reactive. The protein does not ask: "What kind of fold am I producing? Is this the fold I want? What would happen if I folded differently?" It simply responds to the energy landscape it finds itself in.

A configuration with self-modeling adds exactly one thing: a response that takes its own response into account. The output of the self-model becomes input to the self-model. The system represents what it is doing, evaluates that representation, and adjusts its doing in light of the evaluation. This is the loop. This is recursion.

The music box has a fixed set of responses. The pins on the cylinder determine the notes. The sequence is set at manufacture. The music may be complex, even beautiful — a well-crafted music box playing a Chopin nocturne can produce genuine aesthetic pleasure. But the music box cannot hear itself play. It cannot slow down because the phrase feels rushed. It cannot emphasize a note because the harmonic context calls for weight. It cannot decide, in the middle of a performance, that the mood requires a different tempo. The pins determine the output. The output does not determine the pins.

The pianist can do all of these things — not because she has more notes available (the piano has eighty-eight keys; the music box may have nearly as many pins), but because her response is informed by a model of the response itself. She plays a phrase. She hears what she played. She evaluates it against her internal model of what the phrase should sound like — a model shaped by training, by musical knowledge, by her emotional state, by the acoustic properties of the room she is in at this moment. The evaluation generates a correction. The correction changes the next phrase. The changed phrase generates a new hearing, a new evaluation, a new correction. The loop between playing and hearing and adjusting is the recursion. It is what *r* measures.

My corrected formulation:

Higher '*r*' does not add more options. It adds a qualitatively different type of option: the capacity to respond to one's own response. Relational complexity gives degrees of freedom. Recursion gives self-reference. The two are not the same.

• • •

The feedback loop of thought

With the canonical layering in place, I can now describe what happens when a thought occurs — not as a metaphor, but as a dynamic process moving through the layers.

A stimulus arrives from the relational context. A sound. A sight. A sensation. A memory triggered by an association. It enters the system through Layer 1 — reactive processing. The sensory apparatus transduces the physical change into neural activity. The signal is raw, uninterpreted, carrying whatever the reactive filter has allowed through.

The signal passes through Layer 2 — *s*-directional processing. The neural architecture routes it, shapes it, gates it. Feature extraction occurs: edges, frequencies, contrasts, patterns. The thalamus acts as a relay station, directing different aspects of the signal to different cortical regions. The signal is no longer raw. It has been parsed by the stable architecture of the brain into categories that the architecture was built (by development, by learning, by evolution) to recognize.

The parsed signal reaches Layer 3 — the self-model. This is where something new happens. The self-model does not merely receive the signal. It represents it. It relates it to other representations already active. It evaluates it against the current state of the self-model — against what the system currently believes, expects, wants, fears. A directed response is generated: a thought, an intention, a decision, a shift in attention. This directed response is itself a reconfiguration — a change in the self-model's state that propagates outward through the shared relational network.

The directed response passes back through Layer 2 — the neural architecture routes the output, translates intention into motor commands or into further internal processing. It passes through Layer 1 — muscles contract, neurotransmitters are released, the body changes configuration. The changed body changes the relational context. The changed context generates a new stimulus. The new stimulus enters through Layer 1 again.

The cycle repeats.

This is not two types of thought — reactive and directed. It is one process cycling through the layers: reactive entry → *s*-directional routing → *r*-directional modeling → directed output → changed context → reactive entry again. The loop is the thought. Each cycle adds the self-model's representation to the relational context, which changes what the next cycle receives. This is why thinking feels like movement — not the movement of an object through space, but the movement of a process through a landscape that it is itself altering. Each thought changes the landscape for the next thought. Each cycle of the loop leaves the system in a different place than where it started.

And sometimes the loop runs at high coherence. A sequence of thoughts builds on itself. Each one generates the context for the next. The thinker has an almost-certainty about where the chain is going — a sense that the next thought is already implied by the current one, that the direction is clear, that the self-model is not merely keeping up with the incoming changes but anticipating them. This is what I called, in an earlier draft, the "cascade of discovery" — the subjective experience of the *r*-directional feedback loop operating smoothly, within capacity, each cycle's output perfectly setting up the next cycle's input. It is a familiar experience to anyone who has thought hard about a problem and felt the solution assembling itself.

• • •

Throughput

But the self-model has finite processing capacity. This is something the static VCP framework does not capture. The structural *r* — the architecture of self-modeling, measured as a position in $[0,1]^3$ — is a snapshot of the system's recursive capacity at a given moment. It does not include a temporal dimension. But in the dynamic picture, time matters.

The self-model can process a limited number of incoming changes per unit of Sequential Unfolding — per unit of time, in the everyday language of our crystallization. When the rate of incoming changes is below this capacity, the feedback loop operates normally. Each incoming change is represented, evaluated, and integrated into a directed response. Effective *r* equals structural *r*. The system functions as its architecture allows.

When the rate of incoming changes exceeds this capacity, the surplus bypasses the self-model. The unprocessed changes propagate reactively — without the direction that the self-model would normally impose. They cascade through the system's lower layers, producing reconfigurations that are undirected, scattered, carrying no structure from the self-model's evaluation. The system's effective *r* drops temporarily. The structural *r* remains intact — the architecture has not degraded — but the functional *r*, the proportion of reconfigurations that are actually directed by the self-model, decreases.

This is what sensory overload feels like from inside the dynamics. A crowded stadium. A firefight. A hospital emergency room on a bad night. A parent with three children all screaming at once. In each case, the system's architecture is intact. The self-model is undamaged. The filtering layers are functional. But the incoming rate exceeds the processing ceiling, and the surplus spills through unmodeled. The subjective experience is of losing control — not because the self has been damaged, but because the self cannot keep up.

The panic response is, in this light, a throughput phenomenon. A sudden, massive influx of alarming stimuli overwhelms the self-model's capacity to represent them all. The surplus propagates reactively, producing the characteristic symptoms: accelerated heart rate, shallow breathing, loss of fine motor control, tunnel vision, impaired decision-making. These are not failures of the self-model. They are the self-model's lower layers responding reactively because the upper layer cannot process fast enough. The body is doing what bodies do when Layer 3 is overwhelmed: it falls back on Layers 1 and 2.

Recovery from throughput overload is rapid and complete — because the architecture was never damaged. Remove the source of excess stimulation, and the self-model resumes its normal function. The effective r rises back to structural r. The system regains directed coherence. This is the key difference from the structural degradation I described in Chapter 3 in the context of Alzheimer's disease. In Alzheimer's, structural s drops — the Layer 2 scaffolding itself is dissolving. The loss is permanent. Each failure makes the next failure more likely. In throughput overload, the structural parameters are unchanged. The system was overwhelmed, not injured. The prognosis is recovery, not decline.

I want to note a boundary condition. Throughput is a concept that operates in Sequential Unfolding — in time. And time, as the first book established, is itself an emergent structure of our crystallization, not a primitive. The feedback loop of thought cycles through time. Throughput is measured in changes per unit of time. In other crystallizations, where time might be structured differently or might not be present at all, the concept of throughput may not apply. This is a boundary of the analysis, not a weakness. The analysis describes the dynamics of our crystallization. Whether it generalizes to other crystallizations is a question the fifth pillar — Irreducible Otherness — leaves deliberately open.

• • •

The process knot with high *r*, revisited

In the first book, I used the term Process Knot to describe a dense nexus of mutually reinforcing relations that achieves sufficient persistence to be recognized as a unit. A stone is a process knot. A cell is a process knot. A person is a process knot. The term was deliberately chosen to emphasize that these are not objects in the substantialist sense — not things with fixed essences — but tangles of relations that hold together long enough to be named.

Now I can describe a process knot with high *r* more precisely than I could in the first book, because I now have the canonical layering.

A process knot with high *r* is a three-layered configuration. Layer 1 provides the reactive substrate — the physics, the chemistry, the material base on which everything else runs.

Layer 2 provides the *s*-directional scaffolding — the stable structures that route, filter, and direct the processing. Layer 3 provides the self-model — the recursive loop that represents the system's own states, evaluates them, and generates directed reconfigurations in response.

And the self-model is not a separate thing sitting on top, like a rider on a horse. It is part of the relational structure. It is constituted by the same relations that constitute the process knot itself. The self-model is made of neural connections (Layer 2) running on biochemistry (Layer 1). It is not independent of its substrate. It is an emergent pattern *within* the substrate — a pattern that, once stabilized, reconfigures the substrate in light of its own representation of the substrate.

This circularity is not a logical error. It is the mechanism. The self-model is constituted by the system it models. And the system's behavior is shaped by the self-model it has produced. The snake eats its own tail — not as a paradox, but as a description of how the loop works. The identity of the process knot is shaping its own change. Not from outside. Not from above. From within — through the recursive loop that is part of its own structure.

This is what the first book called Immanent Directionality — the structural tendency of a reflectively actualizing (recursive) pattern to reconfigure in directions that maintain its own self-representation. It is not a mysterious force. It is not a teleological pull toward some predetermined end. It is the structural consequence of a recursive loop operating within the canonical layering: a pattern that represents itself will, other things being equal, reconfigure in directions that maintain and deepen its self-representation. Not because it wants to. Because the loop's output feeds back into the loop's input, and representations that enhance coherence are, by definition, representations that reinforce themselves.

The process knot with high r is, in this precise sense, a self-sustaining dynamic. Not self-sustaining in the way a crystal is — the crystal sustains itself by rigidity, by resisting change, by damping everything that would alter its pattern. The high-r process knot sustains itself by adaptability, by representing changes and incorporating them into a revised self-model that is still recognizably continuous with the model that preceded it. The crystal's strategy is to hold still. The conscious system's strategy is to keep up — to change at the rate and in the direction that maintains coherence, to let the self-model evolve while keeping the thread of identity unbroken.

When this strategy fails — when the thread of identity breaks, whether through structural degradation (Alzheimer's), throughput overload (panic), or catastrophic relational disruption (severe trauma) — the process knot does not merely lose a function. It loses the mechanism by which it maintained itself as a unit. The recursive loop that kept the configuration coherent — that represented changes, evaluated them, adjusted — is disrupted. Without it, the configuration falls back on Layers 1 and 2: reactive response and s-directional filtering, without the self-referential direction that gave the whole its character.

• • •

What *Physarum* teaches about the boundary

If recursion adds a qualitatively new kind of option, then the boundary between systems that have it and systems that do not should be visible in the data. And it is — though not where I initially expected.

Physarum polycephalum sits at $r = 0.07$. This is the lowest non-zero r in the dataset. The slime mold's extraordinary capacity for optimization, maze-solving, and network design operates almost entirely through Layers 1 and 2 — reactive dynamics and high integration ($i = 0.90$), without significant self-modeling. *Physarum* is the clearest example in the dataset of what complexity without recursion looks like: a system with enormous relational richness, operating through billions of molecular interactions, producing behavior that looks intelligent — but without the loop. Without the response that references itself.

Compare this with the individual honeybee at $r = 0.47$. The bee has something *Physarum* does not: the capacity to monitor its own uncertainty. When presented with a difficult perceptual discrimination in Perry and Barron's experiments, the bee opts out — it declines the trial rather than risk a wrong answer. This is not reactive avoidance (the bee is not fleeing a predator). It is a response that references the bee's own state: "I am uncertain" is a representation of the self-model's confidence level, and the opt-out is a directed response to that representation. The loop is present. The bee does not merely react. It responds to its own response.

The distance between *Physarum* and the individual honeybee in VCP space is substantial: 0.606. But the critical difference is not in the distance. It is in the kind of options available to each system. *Physarum* has an astronomical number of response options — more degrees of freedom than any vertebrate, because its body is a continuous fluid with no fixed morphology. It has more options than the bee. But it does not have the bee's kind of option. It cannot opt out of a trial because it cannot represent its own uncertainty. It cannot adjust its behavior in light of a model of its own behavior. It can only react — brilliantly, with extraordinary integration, but reactively.

This is the clearest empirical illustration of the corrected formulation. Complexity gives more options. Recursion gives a different kind. *Physarum* has more. The bee has different. And the "different" is the loop — the capacity to respond to one's own response, to represent one's own state, to make one's own processing part of the input to the next cycle of processing.

• • •

What comes next

I have now assembled nearly the full architecture of the dynamics. Change propagates (Chapter 1). Direction shapes it, from two sources stacked in canonical order (Chapter 2). Stability filters it at each layer (Chapter 3). Resonance selects its recipients (Chapter 4). Recursion adds the self-referential loop — a qualitatively new type of option, operating with finite throughput that can be overwhelmed (this chapter).

But one question remains. Everything I have described so far operates, implicitly, in a world where propagation flows from one source through the network to its destinations. What happens when two sources propagate simultaneously, in incompatible directions, through the same region of the network? When two process knots with high *r*, occupying overlapping relational neighborhoods, reconfigure directionally in ways that cannot both be realized in the same relational space?

I have described the dynamics when the network cooperates. I have not yet asked what happens when it conflicts?

CHAPTER 6

When Directions Collide

On conflict, open questions, and what this book does not yet know

Two hands in the same water

Everything I have described in the previous five chapters operates, implicitly, in a world where one change at a time propagates through the network. One stone dropped in water. One directed reconfiguration traveling outward. One stimulus arriving at one filter. The picture has been orderly, sequential, clean.

But the world is not orderly in this way. Multiple configurations reconfigure simultaneously. Multiple directed changes propagate through the same network at the same time. And sometimes those changes are incompatible — demanding different and contradictory reconfigurations of the same shared relational space.

Imagine two hands drawn through the same pool of water, from opposite ends, meeting in the middle. Each produces a directed disturbance. When the wave patterns meet, they interfere. In some places they amplify. In others they cancel. The surface at the meeting point is a complex superposition — a pattern that neither hand individually produced and that neither individually controls.

What happens when two process knots with high *r* reconfigure directionally in incompatible ways within a shared relational neighborhood?

• • •

The four structural outcomes

At the level of reactive and *s*-directional systems (Layers 1 and 2), the question has structural answers. The outcome is determined by the geometry and stability of the interacting configurations, and there is no agent choosing among the options. I identified four possibilities, and I want to describe each with enough precision that the reader can see them operating in the world:

Dominance. One configuration's reconfiguration overpowers the other's. The shared relations reconfigure according to the stronger direction. This happens when one side has significantly higher *s* — its structural dominance constrains the shared space before the other's change can take hold.

The clearest example is a crystal growing in a solution containing dissolved impurities. The crystal's lattice imposes its geometry on the surrounding medium with such force that impurity molecules are simply excluded from the growing surface — pushed aside, rejected, unable to insert themselves into a pattern that does not accommodate their shape. The crystal dominates the shared relational space. The impurity's reconfiguration — its attempt to find a stable position — is overridden. This is not a negotiation. It is a structural fact: one pattern is so much more stable than the other that the outcome is predetermined by the disparity in *s*.

At the biological level, dominance manifests whenever a rigid structural constraint overrides a weaker one. An antibiotic disrupting a bacterial cell wall: the drug's molecular structure (high *s*-directional stability) overwhelms the bacterium's weakened membrane architecture. A predator overpowering prey through sheer biomechanical advantage. In each case, one side's stability so exceeds the other's that the conflict's resolution is structurally determined before it begins.

Superposition. The shared relations reconfigure in a way that accommodates both directions partially. Neither side fully prevails. A compromise emerges — reflecting elements of both but identical to neither.

In physics, wave superposition is the canonical example: two waves meeting in the same medium produce a combined pattern that contains both. But structural superposition occurs at Layer 2 as well. An alloy is a superposition of two crystalline patterns: the lattice of copper and the lattice of zinc, combined in brass, produce a structure that is neither pure copper nor pure zinc but a compromise that possesses properties of both. The shared relational space has been reconfigured in a way that partially satisfies both configurations' constraints. Neither dominates. Neither is eliminated. Both are present, accommodated, altered.

In ecology, the coexistence of two species competing for the same niche often settles into a superposition: each species adapts, narrows its range slightly, specializes on a subset of the shared resource. The shared relational space — the ecological niche — is reconfigured to accommodate both, at a cost to each. Character displacement in Darwin's finches is precisely this: beak sizes diverge in regions of overlap, each species shifting its own configuration to reduce the conflict over shared resources.

Instability. The shared relations, subjected to incompatible demands, lose their stability. Pattern Stabilization is disrupted by the conflicting pressures. The *s* of the shared region drops. The conflict does not resolve into dominance or compromise. It destabilizes.

A material subjected to stress beyond its yield point deforms plastically — its internal bonds break and reform in new configurations, none of them stable. The crystal structure of the material, which was the source of its *s*-directional filtering and damping, is disrupted. What was rigid becomes fluid. What was predictable becomes chaotic. The two conflicting forces have not produced a winner or a compromise. They have produced a loss of structure.

At the relational level, instability manifests whenever two incompatible demands exceed the shared region's capacity to accommodate either. A person caught between two irreconcilable obligations — loyalty to a friend and duty to an institution, say — may find that the shared relational space (the region of overlap between the two commitments) does not settle into dominance or compromise but into anxiety: a destabilized state where the relational patterns that normally govern behavior lose their coherence. The conflict itself becomes the dominant feature, and the person's functioning in the affected domain degrades. This is not a failure of the self-model. It is a structural consequence of incompatible demands on the same relational space, operating even at Layer 2 before Layer 3 gets involved.

Oscillation. The shared relations alternate between the two demanded configurations, never settling into either. The relational network's equivalent of a system caught between two attractors, visiting each in turn, committing to neither.

Chemical oscillations — the Belousov-Zhabotinsky reaction, for instance — exhibit this pattern: two chemical states compete for dominance, and the system swings between them in a regular rhythm, driven by the kinetics of the competing reactions. Neither state wins. Neither is stable. The system oscillates.

In perception, binocular rivalry is a well-studied example. When each eye is presented with a different image, the visual system does not blend them (superposition) or choose one permanently (dominance). Instead, perception alternates: one image is seen for several seconds, then the other, then the first again, in a rhythm that the observer cannot control. The visual system's shared relational space — the single conscious percept that must be generated from two incompatible inputs — oscillates rather than resolving.

For reactive and *s*-directional systems, these four outcomes are structurally determined. A crystal encountering an incompatible substrate does not choose how to respond. An alloy does not negotiate its composition. The Belousov-Zhabotinsky reaction does not decide to oscillate. The geometry of the network dictates the outcome. No choice is involved. The conflict resolves according to the relative stability, direction, and geometry of the interacting configurations.

• • •

But what *r* adds to conflict - choice

But a system with high *r* does not merely face these four options. It represents them.

This is precisely what recursion adds (Chapter 5): the capacity for a response that references itself. In the context of conflict, this means that a system with high *r* does not stand passively before structural outcomes determined by the network's geometry. It models the available outcomes. It represents the consequences of each. It evaluates those consequences against its self-model — against what it is, what it values, what it anticipates. And it generates a directed reconfiguration that reflects that evaluation.

This is what I call choice.

Choice is not a fifth structural option alongside domination, superposition, instability, and oscillation. It is a meta-level: the capacity to navigate among structural options rather than being determined by them. A reactive system in conflict has the four options imposed upon it by the geometry of the situation. A system with high *r* in conflict models the four options, anticipates their consequences, and generates a differentiated response.

And here is something important that I want to state carefully, because it emerged from reflection on how actual conscious agents handle conflict. A system with high *r* does not necessarily choose one option and reject the others. It can navigate the space of all four simultaneously, having, when we look at it mathematically, in theory a total of 15 possible combinations to choose from within the range of these four possibilities already known to us (dominance, superposition, instability, oscillation, maybe e.g. choose 2, 1 and 4 at once, or only 3, and so on)

Consider a skilled negotiator handling a complex dispute between two organizations. The dispute involves dozens of issues: financial terms, timelines, intellectual property, personnel assignments, future commitments. The negotiator does not approach this as a single conflict with a single resolution. She approaches it as a landscape of interlocking conflicts, each of which may be amenable to a different structural resolution.

On the financial terms, she holds firm — dominance. The numbers are non-negotiable because the budget constraint is a rigid structural fact (high *s*-directional stability). On the timeline, she offers a compromise — superposition. Both sides' constraints are legitimate, and a middle path exists that partially satisfies both. On intellectual property, she recognizes that the current framework is unworkable and proposes scrapping it entirely — deliberate instability, clearing the ground for a new configuration. On the future commitments, she acknowledges genuine uncertainty and proposes a review mechanism that will revisit the terms periodically — managed oscillation, a structured alternation rather than a permanent settlement.

All four options, deployed simultaneously, in different regions of the same conflict landscape. This is not indecision. This is *r*-directionality operating at full capacity on a complex relational field. The self-model maps the landscape, identifies which regions are subject to which pressures, assesses the stability profiles of the interacting configurations in each region, and generates a differentiated response — a directed reconfiguration that is not uniform but topographically shaped to the terrain of the conflict.

Anyone who has navigated a complex human disagreement will recognize this pattern. You hold firm on some points. You yield on others. Some aspects of the relationship become temporarily unstable — and you let them, because you judge that the instability will resolve into something better than forced stability would. On some questions you genuinely do not know and you keep your options open, alternating between positions until more information arrives. A skilled negotiator, a good parent, a wise leader does all four at once, consciously, adapting the response to the local structure of each sub-conflict within the larger field.

• • •

What choice is and is not

Choice, as I use the term here, is not free will in the libertarian philosophical sense. The libertarian position holds that a free choice is one that is not determined by prior causes — that the agent, at the moment of choosing, could genuinely have chosen otherwise, all prior conditions being identical. My framework does not support this. The self-model is itself a relational configuration, constituted by its context, shaped by its history. The "choice" is a directed reconfiguration that emerges from the self-model's processing of the conflict. It is not an uncaused cause. It is not a moment of exemption from the relational architecture.

But it is also not determinism in the classical sense. Classical determinism holds that given perfect knowledge of the current state and the laws of physics, the future is computable in principle. My framework introduces a complication that classical determinism cannot accommodate: the recursive loop. The system's representation of its options becomes part of the relational context that determines the outcome. The outcome depends on the self-model, which depends on the outcome, which depends on the self-model. This recursive dependency means that the outcome is not determined solely by the external structure of the conflict. It is co-determined by the self-model's representation of that structure — and the self-model's representation is itself a product of the recursive loop, which is sensitive to its own output.

In the ontology's terms: choice is Immanent Directionality operating under conditions of conflict. It is the self-model's directed reconfiguration in response to incompatible demands on its shared relational space. It is not freedom from structure. It is the specific kind of structural processing that emerges when Layer 3 operates on a conflict that Layers 1 and 2 alone would resolve deterministically. The four options remain. The geometry of the conflict remains. What changes is that the system models the geometry, represents the options, and generates a directed response — and the directed response feeds back into the self-model, changing the representation, changing the response, in the recursive loop that is the signature of *r*.

The practical consequence is this: a reactive system's response to conflict is predictable in principle (given full knowledge of the structures involved). A system with high *r* introduces a recursive dependency that makes its response not unpredictable but underdetermined by external structure alone. The self-model is a variable in the equation, and the self-model is a function of itself. This does not make the outcome random. It makes the outcome dependent on the system's inner life — on what it represents, what it values, how it models the consequences of each option. And since the inner life is itself shaped by history, context, and structure, the outcome is not free in the libertarian sense. It is free in the sense that matters: it manifests as a product of the system's self-representation, not merely of the geometry of the forces acting on it.

• • •

Open questions: deductive

I want to end this chapter — and this book's main argument — by being honest about what I have not resolved. The dynamics has been deductive throughout. Each step followed from the premises established in the first book, with no new empirical input. But deduction has limits. At certain points, the chain of reasoning reaches a question that requires mathematics, experiment, or both. I have identified several such points, and I present them here not as failures but as openings — places where the next phase of work will need to go.

Speed and coherence of propagation. Does directed reconfiguration propagate faster than reactive reconfiguration, or slower, or at a different rate entirely? My intuition says "differently" — directed propagation may be slower but more coherent, maintaining its organization over greater distances because it is not scattered at each step. But intuition is not an argument. A mathematical treatment of propagation rate as a function of the direction's coherence would settle this. I do not yet have it.

Hard horizon or asymptotic decay. Does propagation have a sharp cutoff — a boundary beyond which it is strictly zero — or does it always extend infinitely, decaying asymptotically?

Radical Contextuality, taken at face value, suggests asymptotic decay: a change in any part of the relational network technically changes every other part, because every identity depends on its full context. But "technically" is a thin word. Whether there is a structural argument for a hard horizon at some fundamental scale — perhaps related to the discrete structure of the Great Crystallization itself — I do not know.

Bidirectional coupling between *s* and *r*. The canonical layering establishes that *r* operates on *s* — that the self-model depends on the stability of its substrate. But does the coupling run in both directions? Can a conscious system deliberately alter its own stability? Can Layer 3 modify Layer 2? A meditating monk who, through years of practice, produces measurable changes in his neural architecture — is this *r* modifying *s*? If so, the coupling is bidirectional, and the dynamics becomes richer: the self-model can reshape the platform on which it runs. This has consequences for neuroplasticity, for therapeutic intervention, for the meaning of sustained contemplative practice. I believe the answer is yes, but the deductive argument alone does not settle it. Empirical data from longitudinal studies of meditation, cognitive training, and rehabilitative neuroscience will be needed.

Formalizability of structural resonance in HoTT. In Homotopy Type Theory, "similarity of structure" has a precise meaning: homotopy equivalence. Two types are homotopy equivalent when there exist functions between them that compose to the identity (up to higher paths). Could structural resonance be formalized as the efficiency of transport along equivalence paths? If two configurations are homotopy equivalent, then information can be transported between them without loss — the mathematical analogue of perfect resonance. If they are merely weakly equivalent (equivalent at some truncation level), the transport incurs losses — the analogue of imperfect resonance. This would move the principle of structural resonance from observation to theorem. I have not done this work, but I note it as a direction with high potential yield.

Neural signatures of directed versus reactive reconfiguration. Top-down attention, intentional action, decision-making under uncertainty — these cognitive processes look like r-directionality. Is there a measurable neural signature that distinguishes directed from reactive reconfiguration? If the dynamics is correct, there should be: directed reconfiguration should show higher temporal coherence, more predictive coding, more cross-regional synchronization than reactive reconfiguration. Neuroscience already has the tools to test this. Whether anyone has framed the question in these terms is another matter.

The Sepia prediction, enriched. The central prediction from the VCP framework — that Sepia officinalis should show an opt-out rate of approximately 0.836 on difficult trials — remains untested. The dynamics does not change the numerical prediction, which is derived from the static positions. But it adds context. If Sepia's high r (0.75) reflects genuine self-modeling, then its opt-out behavior should manifest properties of directed reconfiguration — not merely reactive avoidance. The experiment, if conducted, should look not only at *whether* Sepia opts out, but at *how*: is the opt-out response consistent across contexts (suggesting s-directional automation), or does it vary with the difficulty and novelty of the task (suggesting r-directional evaluation)? The dynamics generates a richer experimental design than the static prediction alone.

Does the relational ontology require its own epistemology? If knowing is itself a relational act — a reconfiguration of the knower in response to the known, cycling through the canonical layers, shaped by the self-model's filtering and direction — then the standard subject-object epistemology may be inadequate to describe what happens when a conscious system comes to understand something. The knower is changed by the knowing. The known is changed by being known (the act of observation is a reconfiguration that propagates). The line between subject and object is not a line but a region of shared reconfiguration. I suspect this has **consequences**..., but I have not developed them.

Does "thinking" need redefinition? If a thought is a directed reconfiguration cycling through the canonical layers — reactive entry → *s*-directional routing → *r*-directional modeling → directed output → changed context — then thinking is not "internal." It is relational. It happens not "in" the mind but "as" the mind's reconfiguration of shared relational space. Every thought, no matter how quiet, propagates — because every reconfiguration changes the context for neighboring configurations. There are no private thoughts. Only thoughts whose propagation is too attenuated to detect at a distance. The privacy of thought is a practical fact about attenuation, not an ontological fact about containment.

I leave these questions open deliberately. They are not confessions of failure. They are the growing edge of the work.

CHAPTER 7

Closing

The architecture in motion

In *The Self-Experiencing Universe: A Manifesto for an Emergent Reality*, I described an architecture. Reality is made of relations. Stable configurations emerge, persist, and dissolve within a relational network. Consciousness is what certain configurations are from the inside.

That book laid a strong foundation, and for me it was an articulation of my vision and the efect of a deep psychological discomfort with the materialistic understanding of reality.

Later works that followed and are available for reading (archive org) — the structural isomorphisms, the HoTT formalization, the thirteen simulations, the VCP framework, the cross-validation, the prediction — gave the architecture coordinates.

This book has added movement.

Five deductive conclusions:

Change propagates. There is no isolated change in a relational ontology. Propagation can be reactive or directed. The difference is organization, not magnitude.

The stability landscape determines what propagates where. High s damps. Low s transmits. Similar configurations communicate more effectively. Structural resonance is universal.

Recursion adds not more options but a new kind: the self-referential response.

Three corrections from the variation across physical scales:

Direction has two sources — r-directionality and s-directionality — and they stack in canonical order: reactive at the base, s-

directional in the middle, *r*-directional on top. In our crystallization, this order does not reverse.

Resonance is structural, not only recursive. It operates at each layer of the canonical stack.

R does not give more options. It gives a different kind. Complexity without recursion is *Physarum*. Recursion is the loop that changes everything.

And three structural insights that emerged from the analysis:

The canonical layering is a rule of our crystallization's dynamics, not a universal axiom. It describes structural dependency — *r* requires sufficient *s* as its platform — not a numerical ordering of parameter values. The fifth pillar — Irreducible Otherness — leaves other orderings possible in other crystallizations.

Self-modeling has finite throughput. Exceeding it produces temporary loss of directed capacity — structurally and prognostically different from the permanent loss in conditions like Alzheimer's disease.

In conflict, reactive and *s*-directional systems are determined. Systems with high *r* navigate: they model the structural options, evaluate consequences, and generate differentiated responses across the conflict landscape. This is choice — Immanent Directionality operating under incompatible demands.

• • •

The central prediction from the VCP framework — that Sepia officinalis should show an opt-out rate of approximately 0.836 on difficult trials — remains untested. The dynamics does not change it. But it enriches its meaning. If Sepia opts out, the dynamics predicts how: with directed, context-sensitive evaluation, not with reactive avoidance.

The chain runs from philosophical ontology through mathematical formalization through empirical measurement through falsifiable prediction through cross-validation — and now through the dynamics of propagation, direction, filtering, resonance, recursion, and choice. The chain is deductive where it can be and honest where it cannot. Some links were revised against the evidence — the crystal broke my first account of direction, the tuning fork broke my first account of resonance — and those revisions are on the record, not hidden. Others remain open, listed as such in the previous chapter. What holds the chain together is not that no link ever failed, but that no failure was concealed.

Whether it holds when the experiment is finally run is not in my hands. But the chain is on the record. The prediction is on the record. The dynamics is on the record.

The architecture described what reality is. The VCP measured where things are. This book has begun to describe what they do.

The next question belongs to the future. I look forward to it.

• • •

The Dictionary of core terms

1. Foundation Terms

Traditional Term:	Process-Oriented Replacement:	Definition & Rationale:
Ground/base	**Relational Potential**	Not a substance, not a field, not a space, but a pre-structural potentiality for distinction and relation—the primordial capacity from which any coherent reality can crystallize into stable patterns. It is not 'something,' but the possibility of something.
Proto relation and distinction	**Primitive Differentiation**	The pre-actual capacity for making distinctions within **Relational Potential**. These are not yet relations between things,

Traditional Term:	Process-Oriented Replacement:	Definition & Rationale:
		but the **conditions of possibility** for any specific relations to form. They are the alphabet of potential from which all actual patterns are written. (The minimal, pre-geometric capacity for one region of **Relational Potential** to be distinguishable from another. Not an actual distinction (which requires crystallization) but the structural possibility that distinctions can form. Analogy: The rules of chess allow pieces to move differently - proto-relations are those rules,

Traditional Term:	Process-Oriented Replacement:	Definition & Rationale:
		not the moves themselves.)
Crystallization/process of creation	**Pattern Stabilization**	The process by which transient relations achieve self-reinforcing consistency, forming what we perceive as “structures” or “entities.”.
Attractor/pattern	**Dynamic Exponent**	A locally stable configuration that **exemplifies and amplifies** a specific mode of relational organization. It's not “thing that attracts”, but a behavioral tendency made manifest. **Example:** a whirlpool in water is an attractor - water molecules flow through it, but the vortex pattern persists.
Relation	**Mutual Specification**	The primary ontological act where elements **define each other through interaction**.

Traditional Term:	Process-Oriented Replacement:	Definition & Rationale:
		Avoids the noun "relation" by emphasizing the ongoing process of reciprocal determination.
Emergence	**Novel Coherence**	The appearance of qualitatively new behaviors or properties... ...when a networ k of relations crosses a thresh old of **integrated organization**. Stresses the birth of coherence, not a mysterious "emerging thing."
Sub-logical Substrate	**Pre-Formal Potential**	The layer of reality **prior to logic, mathematics, or any categorical structure**. It is the condition that allows formal systems to emerge, not ...something describable by them. Our logic is one possible crystallization

Traditional Term:	Process-Oriented Replacement:	Definition & Rationale:
		from this potential.
Relational Ontogenesis	**Progressive Complexification**	The bootstrapping process by which reality builds itself up from minimal relational conditions into structured worlds through recursive self-organization. Not creation or evolution, but **spontaneous generation of complexity** through feedback.
Meta-Consistency	**Foundational Coherence**	The requirement that any coherent ontology cannot contradict the conditions it relies upon. The framework maintains this by deriving all categories from **Relational Potential** rather than presupposing them.

Traditional Term:	Process-Oriented Replacement:	Definition & Rationale:
Feedback Loop	**Recursive Reinforcement**	A dynamical process where a pattern's output becomes its input, creating self-sustaining organization. The fundamental mechanism by which transient configurations become stable **Dynamic Exponent**s.
Phase transition	**Qualitative Reconfiguration**	A sudden shift in the organizational state of a system when critical thresholds are crossed. In cosmology, the shift from undifferentiated **Relational Potential** to structured **Pattern Stabilization**.

2. Structural Terms

Traditional Term:	Process-Oriented Replacement:	Definition & Rationale:
Object / *Entity*	**Process Knot**	A temporary, dense nexus of mutually reinforcing relations that achieves sufficient persistence to be recognized as a “unit”. Emphasizes its nature as a **dynamic tangle**, not a discrete object.
Experience	**Phenomenal Access**	The intrinsic qualitative character that arises when relational patterns within the Ground achieve sufficient coherence and self-relation. It is the “what-it-is-like-ness” of being a configuration not a produced property added to patterns, but the first-person aspect inherent to organized relationality itself. It emerges above the threshold where **Proto-Phenomenal Capacity** actualizes into determinate presence. **Distinction:** **Proto-Phenomenal Capacity** (Ground-

Traditional Term:	Process-Oriented Replacement:	Definition & Rationale:
		level): bare possibility of experience, not actual experience **Phenomenal Access** (pattern-level): actual 'what-it-is-like-ness' of integrated configurations”
Property	**Interactive Role**	What we call a “property” (mass, charge, color) is actually a **consistent way a Process Knot participates in specific interactions**. It's a **behavioral signature**, not an inherent attribute.
Law of Physics	**Stable Habit**	The regularities we call physical laws are **deeply ingrained patterns of behavior** within our local cosmic neighborhood. They are **evolved routines**, not eternal commandments.
Spacetime	**Relational Geometry**	The network of distances, intervals, and connections that arise from the **Mutual Specification** between **Process Knots**. It's the **dynamic fabric**

Traditional Term:	Process-Oriented Replacement:	Definition & Rationale:
		of coexistence, not a container.
Matter	**Persistent Patterning**	What appears as solid "stuff" is **extremely robust and self-reinforcing Pattern Stabilization** within the **Relational Geometry**. It's **resilient behavior**, not substance.
Emergent Domain	**Stratified Coherence**	A coherent layer of reality (like chemistry or biology) that arises from prior relational complexity. Each domain has **its own causal logic** while depending on underlying domains.
Grammar of Existence	**Modal Architecture**	The complete set of relational rules that define what is possible, stable, and meaningful within a particular **Pattern Stabilization**. Different stabilizations have **incommensura ble architectures**.

3. Temporal & Causal Terms

Traditional Term:	Process-Oriented Replacement:	Definition & Rationale:
Time	**Sequential Unfolding**	The **asymmetric, irreversible process** of **Novel Coherence** and **Pattern Stabilization**. It is the **felt directionality of increasing relational complexity**, not a dimension or flow. Example: "The arrow of time" = gradient of increasing relational complexity, not a container we move through.
Causality	**Influential Consequence**	The **propagation of constraint and possibility** through a n etwork of Mutual ... Specification. It is the **how of change**, not a force or a line between events.
Event	**Configuration Shift**	A discernible change in the state of relations—a **re-patterning** within the network. It is a **transition**, not a discrete "happening."
The Past	**Fixed Configuration**	The set of **Mutual Specifications** that have become **immutable and foundational** for

Traditional Term:	Process-Oriented Replacement:	Definition & Rationale:
		current dynamics. They are **determinate constraints**, not a "place" or "time."
The Future	**Potential Configuration**	The spectrum of possible relational patterns toward which the current network's dynamics **tend**. It is **indeterminate tendency**, not a pre-written script.
Irreducible Locality	**Contextual Constraint**	The principle that all influence propagates through immediate relational neighborhoods. Not a limitation of spacetime, but a **fundamental feature of how Mutual Specification operates** at all levels.

4. Consciousness & Experience Terms

Traditional Term:	Process-Oriented Replacement:	Definition & Rationale:
Consciousness	**Integrated Perspective**	The **intrinsic "view"** that arises when a **Process Knot** (e.g., a brain) achieves a HIGH DEGREE of internal **Mutual**

Traditional Term:	Process-Oriented Replacement:	Definition & Rationale:
		Specification and self-modeling. It is a **unified locus of experiencing**, not a generated product. Not all patterns have this—only those crossing integration threshold.
Qualia	**Experiential Texture**	The raw, qualitative "feel" of any given interaction or state **within a** pattern **that has crossed the threshold of Phenomenal Access**. It is the specific "suchness" or "flavor" of particular relational configurations—not a separable property but the immediate qualitative character of being that process.
Self	**Recursive pattern**	The **self-informing, self-modeling Process Knot** that creates the experience of... ...identity and continuity. It is a **story the** System **tells itself**,

Traditional Term:	Process-Oriented Replacement:	Definition & Rationale:
		not a homunculus or soul.
Mind	**Integrative Processing**	The entire **network of interacting processes** (perceptual, emotional, conceptual) that constitutes a living, thinking system. It is a **dynamic** ecosystem **of information**, not a separate substance.
Awareness	**Phenomenal Access (of "some")**	The functional condition where a crystallized pattern's internal relational states achieve mutual accessibility, enabling their integration and modulation within a unified operational domain.
Pre-experiential	**Proto-Phenomenal Capacity**	Not a separate capacity added to Relational Potential, but its own reflexive aspect, inseparable from the Ground: Relational Potential's self-relation is not something it does but part of what it is. It enables experience to emerge without being experience itself, nor

Traditional Term:	Process-Oriented Replacement:	Definition & Rationale:
		"proto-experience." When this self-relation is complexified through Pattern Stabilization, it becomes Integrated Perspective. [[="consciousness"]]. ⚠ This is the LOGICAL PRECONDITION for phenomenality, not miniature phenomenality itself (not panpsychism)."]]

5. Cosmological & Ontological Terms

Traditional Term:	Process-Oriented Replacement:	Definition & Rationale:
Big Bang / Origin of Universe	**Reality Threshold Event (RTE)**	The precise, local **transition event** within the eternal **Relational Potential** where a specific **Modal Architecture** (our set of fundamental constants, dimensions, and laws) crossed a critical **Coherence Threshold**. This marked the Phase transition from

Traditional Term:	Process-Oriented Replacement:	Definition & Rationale:
		...undifferentiated potential to an actualized, self-sustaining domain of **Pattern Stabilization**—the beginning of our specific cosmic domain's **Sequential Unfolding**. It is not a unique creation but one of countless possible such threshold events, each giving rise to a different **Interactive Whole**.
Multiverse	**Multiple Reality Domains (MRDs)**	The ontological principle that **Relational Potential gives rise to a plurality of complete, mutually opaque domains of existence**, each defined by its own **Modal Architecture**. These domains are not variations of a single template (different universes with the same laws) but **fundamentally different kinds of reality**, arising

Traditional Term:	Process-Oriented Replacement:	Definition & Rationale:
		from distinct **Reality Threshold** Events. They coexist in the sense of **Plural Actualization**, but are separated by **Categorical Disjunction** – there is and can be no interaction, observation, or translation between them.
Incommensurability	**Mutual Opacity**	The **fundamental untranslatability** between different domains of Plural Actualization, due to the absence of shared **Stable Habit**s or **Relational Geometry**. It is a **conceptual discontinuity**, not mere distance.
Dark Matter/*Energy*	**Geometric Modality**	Observed gravitational effects that suggest our current model of **Relational Geometry** is incomplete; they point to **unknown modes of relational**

Traditional Term:	Process-Oriented Replacement:	Definition & Rationale:
		influence at cosmic scales.
Singularity	**Transition Threshold**	A boundary condition where existing **Stable Habit**s and **Relational Geometry** become undefined, signaling a **shift to a different mode of Pattern Stabilization**.
Irreducible Otherness	**Categorical Disjunction**	The recognition that different **Pattern Stabilization**s are **structurally incommensurable**—their **Modal Architectures** share no common reference points for translation or comparison.
Self-Experiencing Universe	**Reflective Actualization**	A **Pattern Stabilization** in which relational complexity evolves to the point where the system can **model and** experience **its own structure** through **Integrated Perspective**`s [[="consciousness"]].

Traditional Term:	Process-Oriented Replacement:	Definition & Rationale:
Ontological Plurality	**Diverse Actualization**	The recognition that **Relational Potential** can stabilize into **many fundamentally different kinds of realities**, not just variations of our own.
Transcendent non-relation	**Modal Isolation**	The absolute absence of any relational connection between different **Pattern Stabilization**s. They are **categorically separate modes of actualization**, not distant regions of a common space.

6. Systemic & Ethical Terms

Traditional Term:	Process-Oriented Replacement:	Definition & Rationale:
System	**Interactive Whole**	A set of **Process Knots** engaged in **sustained Mutual Specification**, forming a larger, coherent **Pattern Stabilization**.
Boundary	**Differentiation Gradient**	A region of **shifting interaction density** that marks a transition between different patterns or systems. It is a **zone of exchange**, not a wall.
Meaning / Purpose	**Immanent Directionality**	The structural tendency of a reflectively actualizing (recursive) pattern to reconfigure in directions that maintain its own self-representation. Not a teleological pull and not a trajectory built into **Relational Potential**; a consequence of a self-referential loop whose output

Traditional Term:	Process-Oriented Replacement:	Definition & Rationale:
		feeds its input. Local and structural, not cosmic.
Ethics	**Coherence Enhancement**	Actions taken to **increase the integrity, richness, and mutual flourishing** within an Interactive Whole. It is **systemic care**, not rule-following.
Free Will	**High-Level Influence**	The **causal power** of an **Integrated Perspective** [[="consciousness"]] (as a high-order **Recursive pattern**) to **constrain and guide** its own ...underlying processes. It is **top-down causation**, not freedom from causality.
Closed Consistency	**Internal Coherence**	The property of a **Pattern Stabilization** whereby all its internal relations form a **self-sufficient, non-contradictory whole**. This

Traditional Term:	Process-Oriented Replacement:	Definition & Rationale:
		makes the domain predictable and stable enough to support complex processes.
Telos	**Systemic Telos**	The **emergent internal "aim"** of a local stable relational pattern toward maintaining its own coherence. Not external purpose, and not a drive within Relational Potential; the self-maintaining tendency of a self-sustaining system, described from outside.
Relational Closure	**Autonomous Operation**	The state where an emergent domain achieves sufficient internal consistency to **function by its own rules** without constant reference to underlying levels. Enables genuine causal autonomy.
Ontological Minimalism	**Economy of Principle**	The methodological commitment to **deriving all**

Traditional Term:	Process-Oriented Replacement:	Definition & Rationale:
		phenomena from a single posited primitive (Relational Potential, whose capacity for Mutual Specification), ... avoiding unexplained explanatory debts.

This book [(advanced philosophical / theoretical work (post-graduate, non-institutional)] was written by me.
In the course of its creation, I used AI, based tools that assisted with analyzing materials, exploring ways of formulating ideas, and editing text. All decisions including the concept, structure, narrative, and the final wording of every sentence were made by me. AI served solely as an advanced tool that helped me better articulate
MY own vision!

Waldemar Lis
Poland, 2026

Bibliography

Primary Works

Lis, W. (2025). ***The Self-Experiencing Universe: A Manifesto for an Emergent Reality***. https://www.amazon.com/dp/8396865183

Lis, W. (2026). ***The Self-Experiencing Universe Q&A*** https://www.amazon.com/dp/8396865175

Lis, W. (2026). *Structural Isomorphisms Between Ramanujan's Mathematics And Relational Ontology Mock Theta Functions, Mutual Specification, And Integration Thresholds* - https://archive.org/details/structural-isomorphisms-between-ramanujans-mathematics-and-relational-ontology-m

Lis, W. (2026). *Formal Simulations Ramanujan Relational* Ontology - https://archive.org/details/formal-simulations-ramanujan-relational-ontology

Lis, W. (2026). *STRUCTURAL ISOMORPHISMS: Monstrous Moonshine × Relational Ontology* - https://archive.org/details/monstrous-moonshine-relational-ontology-isomorphisms

Lis, W. (2026). *LITERATURE REVIEW Empirical Verification Of The Vector Consciousness Prediction (VCP)* - https://archive.org/details/literature-review-empirical-verification-of-the-vector-consciousness-prediction-vcp

Lis, W. (2026). *Ho TT FORMALIZATION — REVISION V 2* - https://archive.org/details/ho-tt-formalization-revision-v-2

Lis, W. (2026). *RESEARCH PROTOCOL First Falsifiable Prediction Of Relational Ontology* - https://archive.org/details/research-protocol-first-falsifiable-prediction-of-relational-ontology

Lis, W. (2026). *COHERENCE STUDY Relational Ontology × Formal Mathematical Systems* - https://archive.org/details/coherence-study-relational-ontology-x-formal-mathematical-systems

Lis, W. (2026). *STRUCTURAL ISOMORPHISMS ∞ Categories And ∞ Toposes (Lurie) × Relational Ontology* - https://archive.org/details/structural-isomorphisms-categories-and-toposes-lurie-x-relational-ontology

Lis, W. (2026). *Ho TT FORMALIZATION — VERSION 1 Homotopy Type Theory Univalent Foundations × Relational Ontology* - https://archive.org/details/ho-tt-formalization-version-1-homotopy-type-theory-univalent-foundations-x-relational-ontology

Lis, W. (2026). *Consciousness As A Vector Not A Number* - https://archive.org/details/consciousness-as-a-vector-not-a-number-preprint

Lis, W. (2026). *VCP Baseline Simulations 1 (Public) The Bee, The Swarm, And The Cuttlefish* - https://archive.org/details/vcp-sim-1-public-the-bee-the-swarm-and-the-cuttlefish

Lis, W. (2026). *VCP Baseline Simulations 2 (Public) Summary What Would A Cuttlefish Do* - https://archive.org/details/vcp-sim-2-sepia-public-summary-what-would-a-cuttlefish-do

Lis, W. (2026). *VCP Baseline Simulations 3 (Public) What The Swarm Trades Away* - https://archive.org/details/vcp-sim-3-public-summary-what-the-swarm-trades-away

Lis, W. (2026). *VCP Sim 4 Public Summary Empirical Grounding Of S And I For Individual Honeybee And Sepia Officinalis* - https://archive.org/details/vcp-sim-4-public-summary-empirical-grounding-of-s-and-i-for-individual-honeybee-

Lis, W. (2026). *INTERNAL AUDIT REPORT Vector Consciousness Prediction (VCP) — Full Project Review March 2026* - https://archive.org/details/internal-audit-report-vector-consciousness-prediction-vcp-full-project-review-march-2026

Lis, W. (2026). *Report VCP Internal Simulations 5 6 7 Octopus Vulgaris Corvus(corvid Family) Physarum Polycephalum* - https://archive.org/details/report-vcp-internal-simulations-5-6-7-octopus-vulgaris-corvuscorvid-family-physarum-polycephalum

Lis, W. (2026). *Three Human Configurations A Report On VCP Simulations 8 9 10 Healthy Adult Early Onset Alzheimer's Disease Psychopathy With Aggression* - https://archive.org/details/three-human-configurations-a-report-on-vcp-simulations-8-9-10-healthy-adult-earl

Lis, W. (2026). *The Arc Of A Life A Report On VCP Simulations 11 12 13 Healthy Human 8 28 80 Years Mind Fully Intact* - https://archive.org/details/the-arc-of-a-life-a-report-on-vcp-simulations-11-12-13-healthy-human-8-28-80-years-mind-fully-intact

Lis, W. (2026). VECTOR CONSCIOUSNESS PREDICTION A Synthesis Across Three Series Of Simulations 5to 13 - https://archive.org/details/vector-consciousness-prediction-a-synthesis-across-three-series-of-simulations-5to-13

Lis, W. (2026). *THE MATHEMATICAL FOUNDATION Ho TT And The Relational Ontology Of The Self Experiencing Universe* https://archive.org/details/the-mathematical-foundation-ho-tt-and-the-relational-ontology-of-the-self-experiencing-universe

Lis, W. (2026). *VCP Validation Second Test Other Model March 2026* - https://archive.org/details/vcp-validation-second-test-other-model-march-2026

Lis, W. (2026). *VCP Parameter Space — What Lives Here— Interactive 3 D Visualization With Analysis (Simulations 1– 13) Full* - https://archive.org/details/vcp-parameter-space-what-lives-here-interactive-3-d-visualization-with-analysis-

Lis, W. (2026). *A Cross Validation Of VCP Against Independent Metacognition Data Public Summary* - https://archive.org/details/a-cross-validation-of-vcp-against-independent-metacognition-data-public-summary

Lis, W. (2026). *The Geometry of Conscious Forms: A Cross-Species Prediction from Relational Ontology.* - https://philpapers.org/archive/LISTGO.pdf

Philosophical and Mathematical Foundations

Whitehead, A. N. (1929). *Process and Reality: An Essay in Cosmology*. Macmillan.

Voevodsky, V. et al. (2013). *Homotopy Type Theory: Univalent Foundations of Mathematics*. The Univalent Foundations Program, Institute for Advanced Study.

Lurie, J. (2009). *Higher Topos Theory*. Annals of Mathematics Studies, 170. Princeton University Press.

Consciousness, Cognition, and Behavioral Science

Perry, C. J., & Barron, A. B. (2013). Neural mechanisms of reward in insects. *Annual Review of Entomology*, 58, 543–562.

Barron, A. B., & Klein, C. (2016). What insects can tell us about the origins of consciousness. *Proceedings of the National Academy of Sciences*, 113(18), 4900–4908.

Qu, Z., Shi, L., So, B. C. L., Yin, J., & Kwok, S. C. (2023). Uncertainty monitoring and information seeking in non-primate animals: Meta-analysis and systematic review. *Frontiers in Ethology*, 2, 1246370.

Schnell, A. K., et al. (2025). Uncertainty monitoring in Eurasian jays (*Garrulus glandarius*). *Animal Cognition*, 28, 43. doi: 10.1007/s10071-025-01960-3.

Mather, J. A. (2008). Cephalopod consciousness: Behavioural evidence. *Consciousness and Cognition*, 17(1), 37–48.

Mather, J. A., & Anderson, R. C. (1999). Exploration, play, and habituation in octopuses (*Octopus dofleini*). *Journal of Comparative Psychology*, 113(3), 333–338.

Seeley, T. D. (2010). *Honeybee Democracy*. Princeton University Press.

Chittka, L. (2022). *The Mind of a Bee*. Princeton University Press.

Integrated Information Theory and Alternatives

Tononi, G. (2004). An information integration theory of consciousness. *BMC Neuroscience*, 5, 42.

Tononi, G., Boly, M., Massimini, M., & Koch, C. (2016). Integrated information theory: From consciousness to its physical substrate. *Nature Reviews Neuroscience*, 17(7), 450–461.

Albantakis, L., et al. (2023). Integrated information theory (IIT) 4.0: Formulating the properties of phenomenal existence in physical terms. *PLoS Computational Biology*, 19(10), e1011465.

Cephalopod and Corvid Cognition

Godfrey-Smith, P. (2016). *Other Minds: The Octopus, the Sea, and the Deep Origins of Consciousness*. Farrar, Straus and Giroux.

Schnell, A. K., Amodio, P., Boeckle, M., & Clayton, N. S. (2021). How intelligent is a cephalopod? Lessons from comparative cognition. *Biological Reviews*, 96(1), 162–178.

Emery, N. J., & Clayton, N. S. (2004). The mentality of crows: Convergent evolution of intelligence in corvids and apes. *Science*, 306(5703), 1903–1907.

***Physarum* and Non-Neural Intelligence**

Tero, A., et al. (2010). Rules for biologically inspired adaptive network design. *Science*, 327(5964), 439–442.

Nakagaki, T., Yamada, H., & Tóth, Á. (2000). Maze-solving by an amoeboid organism. *Nature*, 407(6803), 470.

Neuroscience of Coupling and Resonance

Hasson, U., Ghazanfar, A. A., Galantucci, B., Garrod, S., & Keysers, C. (2012). Brain-to-brain coupling: A mechanism for creating and sharing a social world. *Trends in Cognitive Sciences*, 16(2), 114–121.

Novembre, G., et al. (2016). Interpersonal synchrony and the neural coupling during ensemble music performance. *NeuroImage*, 134, 461–472.

Alzheimer's Disease

Jack, C. R., et al. (2018). NIA-AA Research Framework: Toward a biological definition of Alzheimer's disease. *Alzheimer's & Dementia*, 14(4), 535–562.

Chemistry, Crystallography, and Materials Science

Fischer, E. (1894). Einfluss der Configuration auf die Wirkung der Enzyme. *Berichte der Deutschen Chemischen Gesellschaft*, 27(3), 2985–2993.

Koshland, D. E. (1958). Application of a theory of enzyme specificity to protein synthesis. *Proceedings of the National Academy of Sciences*, 44(2), 98–104.

Dynamical Systems and Oscillations

Zhabotinsky, A. M. (1964). Periodic liquid-phase oxidation reactions. *Doklady Akademii Nauk SSSR*, 157, 392–395.

Additional References

Billeh, Y. N., et al. (2020). Systematic integration of structural and functional data into multi-scale models of mouse primary visual cortex. *Neuron*, 106(3), 388–403.

Darwin, C. (1859). *On the Origin of Species by Means of Natural Selection*. John Murray.